What Your Astronomy Textbook Won't Tell You

Clear, Savvy Insights for Mastery

NORMAN SPERLING
FOREWORD BY DAVID LEVY

WITH CONTRIBUTIONS BY DAVID MORRISON,
BRAD SCHAEFER, JOHN WESTFALL AND OTHERS

EVERYTHING IN THE UNIVERSE
SAN MATEO, CALIFORNIA

Library of Congress Cataloging-in-Publication Data

What your astronomy textbook won't tell you : clear, savvy insights for mastery / [edited] by Norman Sperling ; foreword by David Levy ; with contributions by David Morrison ... [et al.].
p. cm.
ISBN 0-913399-04-3 (alk. paper)
1. Astronomy. I. Sperling, Norman, 1947- II. Morrison, David, 1940-
QB43.3.W48 2002
520--dc21

2002012952

Table of Contents

5: Galaxies and Cosmology

Appendix

The Author

Norman Sperling is a popular teacher of intro-astro courses in universities around San Francisco Bay. His unceasing effort to make the course clear and interesting led him to develop the narratives, charts and graphs that fill this book.

Sperling has been a planetarium director, an editor of *Sky & Telescope* magazine, and Science Editor of AltaVista.com. He has over 100 articles in print. He is also an inventor: co-designer of the popular Astroscan telescope, and producer of "The Stars Above" starfinder. His company has published books by comet-discoverer David Levy and telescope-designer John Dobson.

Sperling speaks at a wide variety of events. He has long served as an officer or board member of many astronomical and skeptics organizations. He also testifies as an expert witness in court cases where astronomical issues make a difference.

Contributors

Many people contributed to this volume, but by far the greatest help came from my referee, Brad Schaefer, a very sharp astronomer at the University of Texas. Brad enormously improved many of the points I make and tales I tell. Better still, he contributed several excellent ones himself. But best of all, he saved me from publishing goofs. He has my enduring respect and thanks, and he's earned yours, too.

I thank the many contributors who helped enrich this volume.
- My old friend David Levy has written a stirring Foreword.

- Bob Martino, Bill McClain and Jim Craig maintain an excellent FAQ about star-naming scams. I thank them for letting me adapt it for this book.
- David Morrison makes "Bad Words" look good with his perceptive article. A version addressed to faculty has been submitted to a journal.
- My good friend John Westfall always knows a clever angle, and here contributes an item about just that: obliquity. He also suggested improvements to several articles.
- Chris Anderson commented about "quorbits" in the planetarians' newsgroup, and kindly permitted me to adapt his comment here.
- Charles J. Peterson contributes a delightful anecdote about Robert H. Baker.
- My student Heather Leswing kindly permitted me to publish her term paper on horoscope columns.
- Rudi Paul Lindner contributed perceptive and useful comments about old textbooks.
- Alan R. Fisher has greatly contributed to many of the lessons on which these essays were based.

I invite more contributions from anyone who can enrich this book.

Several articles here appeared in earlier forms in periodicals. I thank those magazines and newsletters for giving them their first audiences.

- The Pull of the Planets: *BASIS,* March 1993
- Horoscopes Flunk Test: *BASIS,* June 1993
- Planets Out of Nowhere: *BASIS,* February 1992
- Oxymoronic Astronomy: *Astronomy,* April 1990
- Sperling's 8-Second Law: *Astronomy,* August 1980
- Putting *Worlds in Collision* in its Place: *BASIS,* November 1998
- Was the Loch Ness Monster an Aurora?: *The Planetarian,* December 1994

The Universe According to My Least-Attentive Students

For many years, I've taught college intro-astro courses in the San Francisco Bay area, and graded between 30,000 and 40,000 quizzes, term papers and exams. Most of my students pay attention, read their assignments, and do well on quizzes.

But some needed sleep, some were distracted, and some didn't read their assignments. At test time, some got confused, some waxed creative, and some just didn't get it. My vengeance is to copy those, letter-for-letter, for your amusement.

I keep no record of who wrote each item, or when or where. Only a few papers per thousand contain boners.

I was relieved to find that mine aren't the only students who occasionally blunder. I thank Dale Cruikshank for contributing his students' goofs on a test about the Pillars in the Eagle Nebula. Thanks also to Liam McDaid and Arthur Upgren for contributing some choice boners.

Dedication

To Leith Holloway,
and to the memory of Bob McCracken,
in gratitude for their help and friendship.

1
Beginnings

Foreword
A Beautiful
4-Minute Universe
by David Levy

From December 17, 1965, to a few nights ago, I have spent some 2,800 hours with my eye at the eyepiece, searching the sky for comets. Over all that time I have come to know the sky pretty well, gaining experience with both the different kinds of objects it offers, as well as some insight on how the sky works. Along the road I have also picked up a few new comets, and countless objects that some call FFNs, or, as you will read in the pages to come, "faint fuzzy nothings." These are faint galaxies, clusters, and nebulæ – all kinds of objects that, through a telescope, look like nothing too much.

It's a good idea to see Nature's big picture when we look at the sky. It's too easy to get overwhelmed by details – and while the sky has a lot of them, it does not require us to understand every one in order to get an appreciation of what it's all about. That is what this particular book tries to do – break through the incidentals and really try, through essays, humor, and quotes from ill-prepared students, to get to the essence of what astronomy is. Did you know, for example, that the common clock, with a face and hands, is a derivative of a celestial sphere, one of the oldest astronomical instruments? You'll see how this small story is a beautiful example of the role that astronomy and astronomers play in our daily lives. Astronomy is not something that a few eccentric people practice in their spare time; it is a beautiful enterprise, worthy of our time and attention, that is meant for all of us.

Another discussion touches on one of the most publicized debates in astronomy these days – the question as to whether the ninth planet, Pluto, is large enough to deserve the title of "planet". It's a question I am interested in partly since I knew the planet's discoverer, Clyde Tombaugh, as a good friend. The question should probably never have come up; in any event, as this book suggests, it is more a question of "what a planet is" rather than "should Pluto be one". Pluto *is* one of the 9 major planets in the Solar System. It is far larger than the largest known asteroid. More important, when formed, it was massive enough to condense under its own gravity to form a sphere, instead of an asteroidal body of irregular shape. Pluto has much more in common with the Earth than Jupiter does. Were we to stand on Pluto we would see a night sky like our own, and a single large moon. We couldn't even stand on Jupiter, we would fall right through to its center!

But for me, a personal story makes the question even more interesting, and illustrates astronomy as a human science. Pluto is unique, and its discoverer also was unique. When I first met him in 1963, I could not believe I was in the presence of one of a handful of people who had discovered a new planet. As I got to know him better, I saw a man committed not only to his science but to language and literature. His puns were legendary: he'd start a lecture with "For 50 years I've been a Plutocrat!" and when a balky missile refused to work at White Sands, where he designed the tracking telescopes, he suggested they "fire it." Clyde Tombaugh was one of the last great observers, and never lost his love of observing. A few years before his death, he rebuffed an attempt by the Smithsonian Institution to obtain one of his first telescopes for their collection with the remark that he was still observing with it. Clyde lived in a time where people still froze at their telescopes, rather than spend their time in front of computer screens. In his last years, he sensed that the debate about the status of Pluto was an attempt to take his signature discovery away from him. We've all grown up in a world of 9 major planets, and there is no reason why this should be changed. There is the argument of history: Pluto has been considered a major planet for almost three-quarters of a century. A change would needlessly make things more confusing and more complicated.

Astronomy should be fun and exciting, and there are plenty of real issues to debate. One of the best took place in 1920. On April 26 of that year, Harlow Shapley and Heber Curtis arrived by train in Washington to discuss the state of the Universe. The question: What were the spiral-shaped nebulæ, and how far away were they? From his studies of Cepheid variable stars, Shapley had already established that our galaxy was some 10 times bigger than scientists had previously thought. Because our own galaxy was so big, Shapley reasoned that the spiral nebulæ were related to our own galaxy, and could not lie very far outside it. Heber Curtis took the side of the old school. Our galaxy was far smaller than Shapley's Cepheid measurements had indicated. Since our galaxy is small, those remote spiral-shaped fuzzy patches are comparable in size and nature to our own galaxy, and are probably far away from it.

The audience didn't know it at the time, but it turns out that Curtis won the debate by using the wrong argument. The spiral nebulæ – now known as spiral galaxies – are comparable in size to the Milky Way, and they are at incredibly vast distances from it. (In 1960, I had to deposit a quarter in the observatory piggybank whenever I mistakenly called a galaxy a nebula.) Although Shapley was correct in his argument about the size of our galaxy, he was wrong about the nature and distance of the spirals. Curtis was right about the nature of the spirals, but for the wrong reason.

The size of the Universe is definitely a topic worthy of debate, but like Bannister and Brasher at Oxford University on May 9, 1954, does it matter who won the race that broke the 4-minute mile? It doesn't matter – both broke the record, and it was such a beautiful race! So was April 26, 1920, the date of the Great Debate that helped opened our eyes and minds to the vastness of the Universe. In the pages that follow, you may disagree with some of the opinions, and you may question the author's unorthodox style at times. But through its comments, funny errata from students, examples, and approach, this unique book should help clarify what is really important and beautiful about astronomy.

Make Textbooks Make Sense

They don't always. Textbooks are marvelous tools in a lot of ways, but they have shortcomings, too. Textbooks explain things the best they can, but students sometimes mistake them for "The Truth" instead of "what current evidence indicates".

This book tries to clarify those books. Most items are very short and come in these categories:
- **Oxymorons:** self-contradicting terms

- **What Astronomy Doesn't Know:** usually neglected by textbooks, because they concentrate on telling what we *do* know
- **Reset Mindset:** newer, more meaningful ways to consider the information (some philosophers call these "paradigm shifts")
- **Right Physics, Wrong Planet**: Ideas that turned out to be mistaken, but later proved useful somewhere else
- **Debunking pseudoscience:** A lot is pushed by people who make money from it when you don't know any better

Textbooks neglect the unknown

Textbooks are the epitome of positivism: They concentrate intensely on telling what *is* known, taking rightful pride in having learned so much about such grand and remote topics. In doing so, however, most neglect to even mention the gaps in our knowledge and understanding, even when they're blatant. So this book tries to point out some of the more important gaps.

We'd love to know everything, of course, but our best instruments and our cleverest minds haven't achieved that yet. When we eventually do learn more, there will be even more unknowns beyond. All Science is always that way.

Students are entirely capable of learning that "nobody knows" a certain thing – they may appear startled to hear such a statement in class, but they learn it well enough. Students can also learn where data aren't sufficient to decide something: I often teach that "I'm quite certain that nobody's quite certain."

Some Textbooks
Don't Tell Enough Background

Textbooks are intensely "present-ist", delivering the scientific understanding of today. Many don't tell how

Science came to this point, so readers won't understand the reasons scientists say what they say, nor recognize the same kind of development next time. And books declare what's known right now so positively that some students don't recognize it's merely a framework into which to place the discoveries of the future. So this book provides historical vignettes and perspectives, beginning with a larger perspective on your textbook's place.

I've added some graphs and charts I've put together to clarify relationships.

Your Textbook

Whichever textbook you use, you need to understand its context.

Your textbook contains a lot of features to help you learn the concepts and information. Use the captions, the glossary, the learning objectives, the chapter-end questions, and the further readings, every time they'll help you learn, not only when your prof assigns them.

Your textbook is far more up-to-date, much better illustrated, and far more informative than my introductory-astronomy textbook:

George Abell:
Exploration of the Universe, 1964

I used George Abell's *Exploration of the Universe* in 1965 as a freshman at Michigan State. It was exciting! Not only did it shovel nifty information at me, it conveyed the excitement of research, and the latest perspectives. It even included a few color pictures. (Textbooks didn't get color on every page till the late 1980s. Prices skyrocketed because that's a lot more expensive to prepare and print.)

When I look at Abell's textbook now, however, I cannot help but chuckle. It is so naïve, so ignorant! The pictures look crude, because we have much better technology nowadays. The data are elementary. Spacecraft had only just reached Mars and Venus. Some concepts seem rather strange because we think of those things differently now. There is no mention of background radiation (discovered later that year) or pulsars (they weren't discovered till 1968), and no spacecraft pictures of Jupiter. Computers were huge, clunky, and rare. In so many ways, they didn't understand their clues – they didn't know impact craters pepper the whole Solar System, and they didn't know rings circle all the big planets.

But my text was certainly a good-faith rendition of the astronomy of its era. The fact that it gave me no hint of all that was to come reveals a trait common to most textbooks: they are overly-positive. They concentrate so much on what they DO know that they neglect to point out what they DON'T know.

Abell's book was definitely a big improvement over the previous dominant textbook:

Robert H. Baker: *Astronomy,* 1930

Baker's book went through 10 editions from 1930 clear into the 1970s, a huge span for any textbook. I often checked it out of my city library while in high school, and was surprised it was *not* the one my prof required in college ... surprised, and soon happy. That's because Abell deliberately included astronomy's excitement, and Baker never did. All the data and pictures and understandings of its time are there – the pictures were the very best available – but recited in a dry, declaratory way. That's the kind of person Baker was. Charles J. Peterson relays this story witnessed by a former student of Baker's:

> One day a student approached Baker in his office at the University of Illinois to seek help on a concept which he was having difficulty understanding. Baker reached over to his bookshelf for the

latest edition of his text. He thumbed to the relevant page and proceeded to read the paragraph pertaining to the student's inquiry.

"I don't understand," responded the student.

Baker read the paragraph a second time.

"That's what I don't understand," replied the student.

Baker then read the paragraph for a third time.

"But I still don't understand," lamented the poor student.

Baker returned the volume to the bookshelf and turned to face the student. "I'm sorry, but I can't help you," he said. "I've given it the best shot I can."

Baker's book is a good-faith rendition of the astronomy of its era, but laughable now. It is so naïve, so ignorant! How primitive they were! They didn't know that galaxies were a *big* story. Spacecraft were still science fiction. Computers were undreamt of. And so on. Astronomers back then were just as smart and clever as modern ones, but they had a lot less to go on, and it shows.

Nevertheless, Baker's book marked a major improvement over:

Forest Ray Moulton: *Astronomy,* 1906

Moulton was a leading astronomer of his time, teaming with Thomas C. Chamberlain to propose how the Solar System might have formed as a result of another star coming very close to the Sun.   Though later data disproved the Chamberlain-Moulton theory, it was advanced for its era.

Moulton's book is now a giggle-factory. The writing is not just passive-dull but downright stodgy. The contents are so naïve, so ignorant! This was before radio astronomy, before anyone knew how fusion works.   It's not that much is wrong, but it sure makes you appreciate how much has been learned since then.

Yet it, too, was a good-faith rendition of the astronomy of its era: full of the latest data, and a few recent pictures. And Moulton marched in the forefront of education: his book was also chopped into small sections and marketed for correspondence courses, an early form of "distance learning". Moulton's textbook first appeared in 1906, and remained in print through the edition of 1938.

For all its shortcomings, Moulton's text was a major improvement over the previous dominant text:

Charles A. Young:
A Textbook of General Astronomy for Colleges and Scientific Schools, 1888

Young was a veritable textbook factory. He produced several different levels of text, topped by this full-math version for the most technical students, and cut down successively for non-math college students, high-school students, and, in *Lessons in Astronomy*, for junior-high. Maybe that should have been titled "Lessens" because of how much Young lessened the book. *General Astronomy* went through about 7 editions from 1888 to 1916.

This book tells you what astronomy knew at the time. It is so naïve, so ignorant! This was before most astrophotography, before mountaintop observatories, before anyone understood stellar spectra or how celestial objects evolve. Reading and laughing at an edition of this, which a student had picked up at a flea market, got me started in studying old textbooks. (Thank you, Carin!) Despite how poorly it has aged, it was a good-faith rendition of the astronomy of its age. And, in turn, a major improvement over:

J. Dorman Steele: *A Fourteen Weeks Course in Astronomy,* 1869

Steele was also a textbook-factory. He wrote *A Fourteen Weeks Course in Chemistry, A Fourteen Weeks Course in Natural Philosophy, A Fourteen Weeks Course in Geology* and others. They were illustrated with the latest woodcuts. And they told what astronomy understood back then. It is so naïve, so ignorant! And so awkward! They didn't yet have much stellar spectroscopy. If you read Steele's book now, read it for humor or history, not for modern astronomy. Modern it is NOT! Steele published several editions from 1869 to 1884. But it was a good-faith rendition of the astronomy of its era. And, especially for readability, a huge improvement over:

Sir John Herschel:
Outlines of Astronomy, 1833

For the 90 years from the time the author's father, William Herschel, discovered Uranus in 1781, till John Herschel died in 1871, they were dominant authorities. His is not merely a textbook but a compendium: it is intended to record full information about the entire subject. Practically every astronomer who could read English kept a copy of this book as the first place to check for information. Usually, they could find answers in Herschel. Only if this source failed did they seek another. And yet any student passing intro-astro now should be able to amplify many of the topics. Herschel's book isn't wrong, but it is very fragmentary.

The first edition was an instant hit in 1833, and new editions kept coming, and coming, and coming. John Herschel died 38 years later, but the book *still* stayed in print; the final edition came out in 1905. A 72-year press run! Staggering!

Though this book contains all the *information* you could want, it conveys absolutely no *interest* at all. Even the dullest lecturer is better than this! All the excitement had to come from the reader, because none can be found in the book itself. And, of course, the stilted language further highlights its age. It is so naïve, so ignorant, so turgid! This was before spectroscopy, before the physical nature of most celestial objects could even be described. Yet it was globally-proclaimed as a good-faith rendition of the astronomy of its era. And it was quite an improvement over:

John Bonnycastle: *An Introduction to Astronomy in a Series of Letters from a Preceptor to his Pupil,* **1786**

This text is the earliest to which I've been able to trace the modern arrangement of topics. While things have certainly changed a lot in proportions and details, it seems to have been Bonnycastle whose arrangement was tweaked by succeeding authors to evolve into the common one used today.

This book is hard to read, not only because of its antiquated language, but also because of its antiquated typography: the "s" is a half-crossed "f", "ct" uses a flowery ligature, and so on. The bulk of this book deals with how things move, because almost nothing was known about what they are physically made of. This was before telescopes grew wider than 25 cm. This book is a good-faith rendition of the astronomy of its era. 8 editions of Bonnycastle's book were published in England between 1786 and 1822. It is so naïve, so ignorant! And so hilarious! Yet, in its time, it was a major improvement over:

James Ferguson:
An Easy Introduction to Astronomy, for Young Gentlemen and Ladies: Describing The Figure, Motions, and Dimensions of the Earth; the different Seasons; Gravity and Light; the Solar System; the Transit of Venus, and its Use in Astronomy; the Moon's Motion and Phases; the Eclipses of the Sun and Moon; the Cause of the Ebbing and Flowing of the Sea, &c., **1768**

James Ferguson had a full-size text (said to have interested William Herschel in astronomy) as well as this cut-down version.

This one takes the literary form of a dialog between college-man Neander and his sister Eudosia. Neander is home for term break, and his sister is pumping him for all the neat stuff he learned in his astronomy course. In the middle of page 75, Eudosia sighs.

Neander: Why do you sigh, Eudosia?

Eudosia: Because there is not an university for ladies as well as for gentlemen. Why, Neander, should our sex be kept in total ignorance of any science, which would make us as much better than we are, as it would make us wiser?

Neander: You are far from being singular in this respect. I have the pleasure of being acquainted with many ladies who think as you do. But if fathers would do justice to their daughters, brothers to their sisters, and husbands to their wives, there would be no occasion for an university for the ladies; because, if those could not instruct these themselves, they might find others who could. And the consequence would be, that the ladies would have a rational way of spending their time at home, and would have no taste for the too common and expensive ways of murdering it, by going abroad to card-tables, balls, and plays: and then, how much better wives, mothers, and mistresses they would be, is obvious to the common sense of mankind. – The misfortune is, there are but few men who know these things: and where that is the case, they think the ladies have no business with them; and very absurdly imagine, because they know nothing of science themselves, that it is beyond the reach of women's capacities.

Eudosia: But is there no danger of our sex's become too vain and proud, if they understood these things as well as you do?

Neander: I am surprised to hear you talk so oddly. – Have you forgot what you told me two days ago? namely, that if you had been proud before, the knowledge of Astronomy, you believed, would make you humble?

Proto-feminism, 1768!

Neander's name means "new man". New, because he's going to college, even though he is from the newly risen moneyed commoners. Until his time, to attend either Oxford or Cambridge (the only colleges in England), one had to be a white, male, member of the Church of England, and member either of the nobility or the clergy. By that standard, I suppose that not one single one of my thousands of students would get into college! How about you? Well, they let us all in now. Let's make the best of it while we're here!

The Ferguson book now makes great comedy for its literary form, as well as for its phrasing and scientific contents. It is so naïve, so ignorant! And so hilarious! This was before Uranus was discovered, before gravity was proven to work beyond the Solar System. The first edition was published in England in 1768, and the last in the US in 1819. Yet it was a good-faith rendition of the astronomy of its times, and a major improvement over:

William Whiston: *Astronomical Lectures Read in the Publick Schools at Cambridge,* 1715

Whiston was Isaac Newton's hand-picked successor as Lucasian professor at Cambridge. (Other famous Lucasian professors: early 2000s – Stephen Hawking; 2400s – Cmdr. Data.) Whiston had a varied career worth looking into. This book poses many difficulties for the modern reader: antiquated typography, stilted phrasing, passive dullness, and overwhelming concern with the today-minor issue of sky motions. Whiston published a Latin edition in 1707, his first English edition in 1715, and a second in 1727, the year Newton died. It is so naïve, so ignorant! And so hilarious! This was before achromatic telescopes, before the first *predicted* return of Halley's Comet. While the contents aren't wrong, they barely hint at the main thrusts of modern Science. Yet Whiston's book was, in its turn, a good-faith rendition of the astronomy of its era, and a major improvement over its predecessors ...

Past, Present and Future

You get the point. Astronomy (if not its college textbooks) goes back to early printing, to mediæval manuscripts, to ancient scrolls, to cuneiform clay tablets and hieroglyphic-engraved stone monuments. And because scientific knowledge progresses, each edition ages rather poorly, and after a while serves better as a poor example than a good one.

Your text stands at the front of this long line. It is the modern culmination of all these successive approximations to what astronomers had learned about the universe. It is a good-faith rendition of the astronomy of right now. It tells the best anyone knows. With spacecraft that have gone as far as ours, with telescopes as big as ours, this is what we have learned.

And it won't end with your book! The author is probably already updating it for the next edition. And future authors will publish new ones after that. Some of what it says may be wrong, but since we don't know *which* things, we teach as best we know. Many future discoveries will bring system to current odds-and-ends. Many future discoveries will bring up important aspects scarcely hinted at so far. But we can't teach them, because that stuff hasn't been learned yet.

20 years from now, we'll know a lot better than some of the things in your book. Will you be the *author* of that one? 50 years from now, a better text will outmode *that* one. And 100 years from now, a more-improved version will relegate *that* one to humor. And 1000 years from now, all *those* will look hopelessly naïve, ignorant, and mistaken! And hilarious!

We teach what we know and understand now because that's the best we can do. That's what *your* book tells, in all good faith, however incomplete or mistaken it may turn out to be. Study it well, use it for all it's worth, learn it as the best anyone can do so far, but learn it as a framework into which the improvements of the future can be plugged in.

[The same can be said for all subjects in which knowledge progresses. Learn all of those subjects with the same perspective.]

Publishers who produce books on many subjects tend to bind their astronomical books in black and dark blue. The colors are as somber as men's suits. This may look appropriate to general publishers, but it makes the bookshelves of astronomers quite gloomy. The exceptions: Mars books are often red, and books about the Sun are usually yellow or orange.

How Does Nature Work?
How Does Science Work?

To many people *outside* Science, these questions mean nearly the same thing. To them, Science is mostly about the *results* of Science: what Science has learned about how Nature works. Students should learn how Nature works, and most textbooks dwell on that.

Scientists distinguish between "Nature" and "Science". To most scientists, "Science" is the set of *processes* by which they *investigate* Nature. Students should learn how scientists work, to judge the value of the results. Most textbooks tell too little about that.

Some Scientific Processes
by Brad Schaefer

Most intro-astro teachers want to teach how Science works, but don't always do it, partly because it is not in many textbooks. Students – and the public in general – need to understand that Science really works in many aspects.

How Science Earns Confidence

Most fundamental is the "scientific method". Students often get a weak linear version in high school ... and it isn't mentioned again, even though it is at the core of what teachers are trying to teach. The linear version from high school is something like: hypothesize-experiment-analyze-conclude and then stop with a smile on your face.

But scientific procedure really never ends, involving tests, retests, double checks, pushing ideas to their limits, and repeatedly checking them out in new settings.

Modern Science never *proves* anything in the way that math constructs Euclidean proofs, since the real world is always more complex than math. So Science makes repeated tests of claims. If the predicted result is found, we get more confidence that the basic claim is correct. The stricter the test or the more unique the prediction, the more confidence a test gives. So experiments and observations are designed to test a claim, and a successful prediction yields more confidence that a claim is true. A failed prediction means that the confidence in the claim is lessened, perhaps critically killed.

The scientific method is more like this:

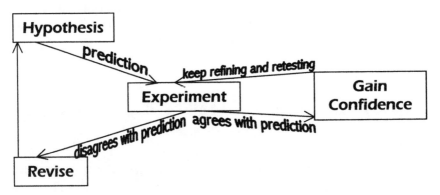

This is much more realistic.   The cycle never ends. Recently, I tested a fundamental hypothesis to 1 part in 10^{21}, and I gained a lot of confidence, but I am now planning to improve on this by a factor of 1000.   This feedback loop makes Science the best way to find truth in an uncertain world.

Political and economic theories largely lack such rigorous testing feedback (even if they had, I wonder how much their adherents would change their views), so mistakes can live on a long time.   There can be bogus or unlucky results in Science also, but inevitably, re-checks weed out the mistakes and lead to greater truths – preferably, quickly.

This loop also shows that all that Science can hope for is to depict reality more and more confidently and accurately. So a theory will always be a 'theory' without the absoluteness of a Euclidean proof.   Nevertheless, it might be such a confident result that we are willing to bet our lives on it.

Students need to evaluate the confidence level of claims based on evidence. How probable is the claim to be true?

For example, the latest black hole theory might appear in the newspapers when it is only 50% likely to be true. The Special Theory of Relativity is 'only' a 'theory', but it has been tested so many times and so critically that it is something like 99.99999999% likely to be right. On the other hand, the *General* Theory of Relativity has not been tested so completely and there are rational alternatives, so most experts would express something like 99.9% confidence that the General Theory is correct.

The idea that 'asteroid impacts resulted in killing off the dinosaurs' went from about a 50% confidence level at the time it was first proposed by Luis Alvarez, up to 95% when the doubly-shocked quartz was found in the Cretaceous-Tertiary boundary layer, to 98% when the Chixulub crater was found in Yucatan, to its current confidence level of about 99.9%.

The hypothesis that the stars really rule human life as in astrology stands no better than the 0.00001% confidence level, having failed every scientific test. The hypothesis that Santa Claus brings gifts to kids every Christmas has much more evidence and thus might enjoy a confidence level of 0.0001%. Big Foot is a hypothesis with perhaps a 0.1% chance, ELF waves cause cancer at the 0.1% confidence level, and pyramid shapes cause the preservation of mummies at the 0.0001% confidence level. And so on. Scientific claims come with a wide range in confidences. Quacks, politicians, and pundits bombard us with claims that have an even wider range.

Error Bars

Nowhere but in Science do you find "error bars", which tell the range of numbers that reality is most likely to fall within. These never appear in newspapers, yet reckoning uncertainties lies at the heart of Science. We simply must be able to evaluate how well a result agrees with the prediction.

For example, the *New York Times* once published a measure of the Hubble Constant to be 79 km/sec/Mpc, at which the advocates of such a high number exulted 'we-told-you-so'. But the newspaper neglected the huge error bar, about ±35, so the result was too poor to distinguish whether the Hubble Constant is really high or low. The *New York Times* really should not have mentioned the measurement at all because it was so poor.

Most error bars mark 68% confidence intervals, meaning that a measured result will come up in the quoted interval roughly 2/3 of the time that an experiment is run. The more hypothesis and measurement disagree, the less likely the hypothesis is to be true.

Textbook science usually has a very high confidence level – generally above 99% – while newspapers average a 70% confidence level. [I can provide details of my measurement of this number.] The difference is caused by newspapers reporting front-line science when it is "news" – the newest information – before there has been time for tests, retests and double checks. Front-line science hasn't had time to become certain, hence its ~70% confidence. Textbook science has had plenty of time to be checked many times and to pass all tests, so it is nearly 100% confident. When textbooks contain less-certain information, they should say so. The public must realize that science claims in the newspapers are far from certain.

Significant Figures

One way to decide how many decimal places an answer should use is to distinguish between *precision* and *accuracy*. *Precision* is how *exact* the answer is presented. *Accuracy* is how *close* the answer is to reality. For example, for п:

- Both very accurate and very precise: 3.141592654
- Very precise but not very accurate: 3.152987650
- Accurate but not very precise: 3.14
- Neither very accurate nor very precise: 3.08

Generally, people do (and should) give answers only as precisely as their accuracy. Giving an answer with more "significant figures" than justified misleads people to think that the answer is more accurate than it really is.

Scientists are Humans, Who Have Motivations

Motivations worm their way into scientific method. Science doesn't formally consider motives, so, in theory, Science is impersonal. Yet in real life, scientists are always checking things out for a wide range of emotions and ideas. For example:

- Trofim Lysenko wanted genetics to be Marxist, so he concocted his new hypothesis based on the Marxist dialectic. Lysenko was being a good scientist to formulate a hypothesis of genetics and then run experiments to test this. [The subsequent cover-ups and death squad hits were politics, not Science.]

- Gerard de Vaucouleurs wanted to disprove Allan Sandage's low measure of the Hubble Constant, so he sought evidence for a high Hubble Constant. His experiments and results were equally valid as if he had been dispassionate. See Dennis Overbye's book *Lonely Hearts of the Cosmos.*

Many times I have tried out experiments when I spotted a flaw in an earlier work, just as other people seek flaws in my work. Getting a famous result can even earn tenure. But by far the most frequent motivation is someone wondering "what if ...".

Motives of scientists are as good and bad as the motives of anyone else. But that's ultimately irrelevant to Science. At most, they can merely speed or slow progress. Motivation can become bad if it induces a blindness in interpreting results incorrectly (a common and very human failing). In Science, the ultimate arbiter is always Nature. Reality will not agree on a fine test with a bad hypothesis, no matter how good the motive. This is so common there's a standard phrase for it: "A beautiful theory, killed by an ugly fact." And a badly-motivated idea that agrees with all experiments represents reality well.

Dead Ends

A whole set of topics that you never hear about in textbooks gives insight into how Science really works. They never tell about all the dead ends.

- In the 1910s and '20s, Adriaan Van Maanen measured what he thought was proper motion in what we now know to be galaxies. He was mistaken, his numbers (in hindsight) impossible, and his work led to nothing.
- Throughout the 1950s and '60s, Fred Hoyle propounded the "Steady State" theory of cosmic evolution, prompting some penetrating observations ... which, however,

support the rival "Big Bang" theories. The Steady State does not appear to describe how the Universe works, though its philosophical implications would have been interesting.

- Arnold Sommerfeld worked in 1916 on modifying Niels Bohr's concept of the atom, giving electrons elliptical orbits! That, too, isn't how Nature works, and didn't lead to anything.

The conceit of hindsight lets you snicker at some dead ends, but keep in mind that the concepts were fair to propose until contrary observations were made. [Ed note: Nature doesn't come with a user's manual. The task of Science is to figure out how Nature works.] There are many dead ends in Science, but textbooks don't show them, concentrating instead on the successes. Some of the cutting-edge science you read about in the media will turn out to be dead ends.

Exploratory Science

Many descriptions of "how Science works" also neglect another major process, even though textbooks abound in examples. A lot of scientists explore. They poke around Nature and see what there is to notice. Astronomers conduct full-sky surveys and seek new phenomena. They send spacecraft to distant objects for close-up looks. They look in different wavelengths. They seek fast-changing phenomena. They seek different-looking objects. Lots of results fill your textbook, and lots of narratives fill books in your library.

The curious thing is that exploration is often left out when describing "scientific methods". In fact, the famous philosopher of Science, Thomas Kuhn, went out of his way to deny it in his book *The Structure of Scientific Revolutions*. Kuhn and others are wrong to deny exploratory science – Science explores Nature all the time.

Coming to Terms With Astronomy

Lots of these essays and notes deal with the words used in astronomy. That's because students have more trouble with the words than with the concepts. Often, the terms are odd, or antiquated, or wrongheaded, or ambiguous, or an unfamiliar definition of a familiar word. When I avoid such troublesome terms, and explain the concepts in plain English, practically all the students understand. It is the terminology that gets in the way of understanding. I can't banish the words, but I can alert students to them. If those terms no longer confuse people, the actual information is much easier to learn.

The Rule of 3 Strange Terms

Why do the terms overwhelm the understandings? I've noticed a principle that I call "The Rule of 3 Strange Terms". Learning something new often comes with new terminology. When presented with the first new term, people learn its definition. When presented with the second, they learn its definition, too. When presented with the third, most people can learn that definition as well, but that's as many new special meanings as most people can handle at one time. Before they can learn more, they have to internalize these so that they're not special meanings to keep in short-term memory any more. If confronted with a fourth new term, instead of learning it, they go on "overload", and dump *all* the new terms, and the understandings that came with them. Reducing new terminology clears the way for more understanding. Math students, especially, would benefit from fewer special terms.

Science Names Phenomena Too Soon, and Clings to Outmoded Names Too Strongly

Some of the terms are strange because they come from Greek or Latin, which very few English-speaking people know. But many are strange because they don't describe the things they label.

The process begins when a new phenomenon is named, shortly after it is first noticed. Then, scientists work on it. After a while, the phenomenon's nature becomes better understood. Very often, it isn't what it was first guessed to be, and named for. The problem comes from keeping the original name after we know it's not really appropriate. The result is lots of phenomena that Science can explain, but which carry names that aren't valid, and that sometimes outright contradict present understanding. Those names get in the way of understanding how Nature works.

A lot of the terms come with tales attached. Those illuminate how Science progresses. They might not fit in a textbook, but they give a much richer view of Science.

12 Bad Words

By David Morrison
NASA Astrobiology Institute
NASA Ames Research Center

Adapted for students from a submission to
Astronomy Education Review (2002)

Almost everyone encounters the problems of technical jargon. Teachers and public presenters learn to avoid most technical terms, or define them carefully; diligent students ask what they mean, or look them up.

But familiar words used in unfamiliar ways endanger communication even more, because listeners and readers may think they understand the term when they do not. Students already know the common meanings of these 12 words. However, their meanings are very different in astronomy and other sciences, and that causes misunderstandings.

believe

The dictionary definitions deal with faith or having convictions, especially religious convictions. [Ed. note: In Science, it means "current evidence indicates".] When scientists "believe" a scientific result or theory (see "theory" below), they're not associating Science with a religious or philosophical belief system. They mean "I think" or "I conclude" or "I understand a scientific point", rather than "I believe it".

body

In everyday English the term relates to an organism or living thing, often someone who is dead. In astronomy, it refers to objects in space, such as comets or asteroids. Most people think this sounds silly. And believe me, explaining that we mean a "heavenly body" does not help!

brightness

In my experience, most people understand "brightness" to mean something like surface-brightness. They also sometimes use the word "color" to mean surface-brightness or reflectance, as in describing a carbonaceous meteorite or the lunar surface as having a "dark color". Astronomers use "albedo", but hardly any students understand that jargon. Less formal terms "reflectance" or "reflectivity" are better but still unfamiliar. More words make it clearer: "what percent of light that a surface reflects."

dense

One common meaning of "dense" is stupid, ignorant, impenetrable – which students may associate with their textbook. In common language, the terms "light" and "heavy" characterize density. Students call an iron meteorite "heavy", not "dense". A puff pastry is described as "light", not of low density. In Science, density means "mass divided by volume". Something dense, like iron, can be light (if it's teensy); something not very dense, like styrofoam, can be heavy (if there's a huge volume of it).

finite

For the public, the opposite of "finite" is "infinite", implying "infinitely large". In some scientific contexts, the opposite of "finite" is "infinitesimal", meaning "infinitely small". The mass of a neutrino may be "finite", as opposed to zero.

limb

Ordinary English firmly ensconces "limb" with two meanings: an arm or leg, or a tree branch. The "limb of the Sun" or the "limb of a planet" means the "zone just inside the edge of the disc" we see. In "limb darkening", the zone just inside the edge looks fainter than the rest of the disc.

magnitude

The ordinary meanings of size or quantity will never lead a student to understand the astronomical magnitude scale. It's easy to learn that astronomers measure how bright something looks; the hard part is the system they use. It reverses common usage, with small magnitudes indicating large brightness.

Science and math also use an "order of magnitude", which means a ratio of 10. One order of magnitude means 10 times bigger or smaller.

mean

Everyone knows the common adjective ("poor", "shabby", "stingy", "base"), but hardly anyone outside of Science knows the mathematical definition: "average". Most people understand "average" but not "mean" or "median". To most people, a "mean value" is a "small or stingy amount"; in Science and math, a "mean value" is the average. [Ed. note: It is possible to be both at the same time. The average instructor's wages are stingy.]

model

A "model" can be a "miniature representation" (such as a model airplane), a "prototype" or "pattern", a "design" (as in a "model year" for cars), a "person who displays clothes", or "someone who poses for a portrait". None of these familiar meanings corresponds to the concept of a scientific model. A scientific model may be ideas and numbers. It doesn't have to be constructed physically; nowadays models tend to be abstract and computationally intensive, like a model of a stellar atmosphere. A scientific model must explain observations, and accurately predict new ones.

radiation

This is one of the most often misunderstood terms, even among different sciences. To the public, the media, and most scientists, "radiation" is short for "ionizing radiation" or "high-energy radiation", usually associated with radioactivity. Except to treat cancer, ionizing radiation is harmful and should be avoided. But to astronomers, *all* light (most of it harmlessly low-power) is "radiation" because it "radiates" from its source. Astronomers study the (harmless low-power) radiation that their telescopes collect. No wonder that farmers have objected to nearby radio telescopes, collecting radiation they think is dangerous! The public, many reporters, and some health-care workers think that cellular phones (which use low-power radio waves) are radio*active*, and worry about possible health effects from their radiation. I have heard a NASA astronomer and a NASA biologist speak together about "solar radiation", neither of them realizing that the astronomer was talking about "light from the Sun", and the biologist about "solar cosmic rays"!

reduce

In English this word means to "decrease" or "make smaller". What, then, is a lay audience to make of an astronomer who speaks of "reducing" data or "reducing" observations? People may think we throw away some of the data. This term is outdated, since modern processing and calibration are very different from the 1800s, when they "reduced" observations to a small number of averaged data points before performing laborious calculations.

theory

Widely misunderstood, this word has multiple meanings for both the public and scientists. It derives from the Greek for "beholding" or "speculation". The general meaning is "speculation" or "guess". I recently read a journalist's report, explicitly writing "My theory is that ... . If this hunch is correct ... ." In contrast, the scientific meaning is "an acceptable general principle to explain phenomena", "a unifying concept that ties together many observations or experiments and has been widely tested and validated". The scientific meaning includes major, rigorous testing that is poles apart from the common meaning; hence the confusion that leads to the common mistaken assertion that "evolution is just a theory". Less frequently, relativity also suffers from the "just a theory" criticism. Students and the public need to understand that gravitation, electromagnetism, relativity, and evolution have been rigorously tested and are far more robust than mere guesswork or hunches.

[Ed. note: for similar terms, see "The Meanings of 'Metals'", "Fusion Confusion: A Burning Problem", and "The Meanings of the Milky Way".]

2
Sky Motions

Sky Motions
according to my least-attentive students

Humans started off with using stones, like stone henge, to map where the stones would come up at the horrizon or at night.

[At Stonehenge] many of the rocks line up on the Summer Solstice.

The astronemers were able to recognize sun set and sun rise by the way it looked on the rocks. The reason why astronemers watched the sky is because they wanted to know how often the sky moved and for how long it was moving.

During the summer the sun rises overhead.

The sun rose and fell at the same time day after day.

The days were longer and lighter on one side of the universe while they were shorter and darker on the other side.

For the Egyptians, twelve moths were counted as one year.

A student of Liam McDaid called the Babylonians "Baby onions".

It was first thought that the stars revolved around earth but of course the opposite is true.

By watching how the ski moves, people can measure things such as time. ... People could also learn about distance such as north, south, east and west by looking in the ski. Seasons could also be measured by the movement of the ski.

By finding out the coordinates of the stars, people have been able to circumvent the planet.

The sky's movement was also beneficial to mariners, who navigated by it, creating tools like sextets and octets.

From Celestial Spheres to Armillaries, Astrolabes, and Clocks

Many books and lectures gloss over attractive old instruments without telling how they relate to one another.

One of the earliest forms was the *celestial sphere*. Pretend that the sky all around you, outside, is a dome very far away. That is the upper half of the celestial sphere. The stars are spots on the sphere. Astronomers chart coordinates to reckon positions on it: right ascension and declination, analogous to longitude and latitude on Earth.

A famous celestial sphere from ancient times is the Farnese Atlas. The Gottorp globe (1664) and the Atwood Sphere (1912) were perforated so people in the dark inside would see "stars" lit from behind – ancestors of the projection planetarium. Since the 1800s, paper-on-wood globes have been sold all around the world. Since the late 1900s, beautiful acrylic globes have been widely used.

On each, sky-mappers not only charted the stars, but the important circles of the celestial sphere. Centuries ago, before the nature and evolution of celestial objects bumped them out of textbooks, students used to have to learn them: right ascension, declination, equator, ecliptic, tropics, Arctic and Antarctic circles, meridian, horizon, and the equinoctial and solsticial colures. Each is defined precisely. Those circles were very useful, but the entire sphere was difficult to make, so they invented another instrument.

Cut away the <u>sphere</u>, and leave only the important circles, to make an <u>armillary</u>

The important circles, minus the sphere itself, constitute an *armillary*, an instrument for measuring positions on the

Celestial Sphere

Armillary Sphere

Astrolabe

Astronomical Clock

Planisphere

Sextant

celestial sphere, though the sphere itself is cut away. Ancient Greeks such as Hipparchus and Ptolemy used armillaries. A few modern decorative versions have degenerated to unusability because the angles are wrong.

Armillaries are quite useful, but also quite bulky, heavy, and expensive. So they invented a simpler instrument.

Flatten an __armillary__ to make an __astrolabe__

Astrolabes were mentioned in ancient writings, and many elegant brass examples survive from the middle ages. Most were used in the Middle East and Mediterranean.

Each pivots around the sky's North Pole. Most have a circle around that for the equator, and another, off-center, for the ecliptic. Each has a decorative way to point out important stars. Another disc (different for each latitude) portrays the horizon, zenith, and altitudes.

A modern descendent is the familiar mass-market cardboard "star-wheel" or planisphere. A lot more people can afford them, but they lack the elegance of old engraved brass.

For compact, light-weight use aboard ships, the full circle was sliced up. A quarter-circle is a quadrant, $1/6$ of a circle is a sextant, and $1/8$ of a circle is an octant.

Motorize an __astrolabe__ to make a __clock__

An astrolabe, with a Sun-pointer, and a motor to make that move at the rate at which the heavens turn, is a clock. The old clock in Prague *looks* like an astrolabe because it *is* an astrolabe. The Sun-pointer became the hour-hand.

Years later, clocks split the 24-hour cycle into two 12-hour rounds, and recently digital clocks have replaced the round face and hands that clocks wore for centuries. But the round clock with hands began as an astrolabe with a motorized Sun-pointer.

All these astronomical devices are directly related. All depend on things moving around the center of the device, representing the Earth at the center of the sky. And we see their descendents around us today.

Wondering about Wandering: How <u>Im</u>proving Astrology Meant <u>Dis</u>proving Astrology, and Made Astronomy a Science

Ancient skywatchers saw 3,000 points of light all marching in lockstep around the sky. And they noticed 7 lights moving differently: the Sun, the Moon, Mercury, Venus, Mars, Jupiter and Saturn. These moved in non-obvious ways, and that attracted attention.

These 7 "wandering" lights seemed to have wills of their own, so the ancient Greeks regarded them as gods. But the ancient Greeks didn't speak modern English, they spoke ancient Greek, in which "wanderers" is *"planetes"*. That's where we get our modern word "planets".

The wanderers come in three obvious sets:

The **Daylight Discs** – the Sun and the Moon – are the brightest by far, and the only ones that don't look like mere points. They always move from west to east against the background of "fixed" stars. And, until careful motion measurements about 2,400 years ago, that motion was thought to be smooth, constant, and circular.

Mercury and Venus are bright points of light. Most of the time, they move west-to-east against the background of stars. But then they slow down, stop, and move east-to-west, crossing the Sun from the dusk sky to the dawn sky. Then they slow down again, stop again, and then resume their usual west-to-east motion. This backward "retrograde" motion looked weird, and it attracted attention. So did the odd fact that Mercury and Venus _always_ stay close to the Sun. They're _never_ opposite the Sun in the sky.

Mars, Jupiter, and Saturn are also points of light. They, too, move west-to-east against the background of stars most of the time. And they, too, slow down, stop, and move east-to-west for a while, then slow, halt, and resume their usual motion. But these wanderers can stray very far from the Sun. In fact, it is when they are opposite the Sun that they go retrograde. These traits were also very odd, and attracted attention.

All these would be minor curiosities if the Sun didn't have so much influence on us.

All peoples have always agreed that the Sun dominates life on Earth. It gives us our light and heat, and without it we couldn't live.

The second-brightest "wanderer", too, keeps pace with some human affairs, as well as the tides.

It was perfectly fair to suggest that if the brightest "wanderer", the Sun, had major influence on us, and the second-brightest, the Moon, also had influence (though less), then the fainter wanderers might have their own effects on us, though more subtle. This began astrology. Before the nature of the planets was known, and before the forces of Nature were detailed with Renaissance precision, this was fair to suggest.

Not only fair, but significant. If the wanderers were gods, influencing humans, then tracking their motions was important for understanding what was happening to humans.   And predicting their motions could enable humans to better cope with coming situations.

From Ptolemy through Kepler to Newton, Western culture's incentive to understand planet motions was to predict planet positions in the sky, to enhance coping with their influences. In one of the great ironies of history, figuring out how they moved led directly to understanding gravity and the other forces of Nature, which in turn demonstrated that the planets do *not* influence humans after all! Trying to improve astrological skill, they disproved astrology, and set up modern astronomy as a Science instead.

Debunk
The Pull of the Planets

The most pervasive form of pseudoscience in our culture may be the syndicated daily newspaper horoscope column. About a dozen of them are published every day.   Even though many professional astrologers denounce newspaper horoscopes, they're probably the "hook" that pulls a lot of believers into the astrologers' webs.   While numerous scientific studies demonstrate how invalid astrology is, none addresses this most common form. A field so full of fertilizer deserves to be sniffed around.

Most horoscope writers are female, or at least their pen names are.

The columnist generally receives half of the column's gross income, but the price paid by each newspaper is not set. It varies with circulation, size of market or territory, and the bargain made between the newspaper and the syndicate. (Weiner, p. 24) In 1992, reported charges ranged from $1 to $15/week, with most between $2 and $4/week. Many horoscopes are packaged with other material, and a newspaper buys the whole package. A successful column will appear in over 100 papers. (Weiner, p. 26)

Syndicated columns were identified in the *Editor & Publisher* annual listing, and *The Working Press of the Nation.* Syndicates are notoriously reluctant to reveal their circulation numbers. Most syndicates ignored, or only partly answered, the inquiry letters I sent them in January 1992. The best and most revealing response came from Naida Dickson of Dickson Feature Service, Gardena, California. I asked what system of astrology she used: "None. Totally random." What is superior or distinctive? "A very glib lady did it." What are Diana Dee's qualifications? "Just a pseudonym (anagram) for Naida D." Statistics on accuracy? "None." Testimonials? "Some [newspapers] used it for months and never paid. For a while, I thought I should offer every possible kind of feature, so I whipped up *Your Lucky Stars.* Now I stick to my original offering: Puzzles only."

When Did Your Cancer Start?

The 11 published horoscope columns examined don't agree about which dates each sign starts on. The only thing they agree upon is that Aries begins on March 21. For all other signs, there is disagreement, in some cases spreading not just over 2 days, but 3. Sonia McGinnis begins her Leo on July 22nd, though Larry White doesn't start his until July 24th. Both use those same day numbers to begin Virgo in August. Capricorn starts on December 21st for Wanda Perry, but not till December 23rd for B. J. Crowley.

So, when did Cancer start? The Cancers of Wanda Perry, B. J. Crowley, Stella Wilder, C. C. Clark, Jeane Dixon, Sydney Omarr, and Bernice Bede Osol all started on June 21st. The Cancers of Sonia McGinnis, Larry White, Selma Glasser, and Joyce Jillson started the next day, June 22nd.

Reference

Weiner, Richard. *Syndicated Columnists.* 3d ed., 1979. New York: Weiner.

A Certain Week

I asked each syndicate to provide their horoscopes for the week of October 13-19, 1991, because of specific real planet conjunctions. The Moon occulted Uranus on October 14; the Moon occulted Neptune on the 15th, when the Moon was also at First Quarter and apogee, and Mercury was at descending node; on the 16th, Saturn was just south of the Moon; and on the 17th, Venus was just south of Jupiter. If horoscopes listed those, fine; if not, whoops; if others instead, double-whoops.

While the syndicates didn't send the requested columns, we found many in newspapers. My student Heather Leswing analyzed them:

Debunk

Horoscopes: Believe Them or Not

By Heather Leswing

I read a total of 210 statements (there were very few predictions) from the daily columns of Jeane Dixon, Sydney Omarr, Joyce Jillson, and Bernice Bede Osol, plus the weekly columns of Larry White and Sonia McGinniss.

References to Celestial Objects

There were just 12 mentions in a week with several significant conjunctions. 8 of the references were made by Sydney Omarr, all of them about the Moon. The other 4 references were made by Cosmic College, 2 of them about the movements of planets and 2 were unspecific.   None referred to any of the actual conjunctions.

Corresponding and Contradictory Statements

23 times, 2 astrologers made corresponding statements on the same day. For Libra on October 14, Dixon, Omarr and Osol all refer to restoring harmony in domestic situations. Considering that over 600 statements were made in that week (most entries say about 3 things), 23 sets of corresponding statements aren't very many.

While reading the statements, I felt at times like each astrologer has a big hat filled with random statements. Every morning they pick our 2 or 3 statements and send them to the syndicate.

If the corresponding statements are just the luck of the draw, then one would think that there would be contradictory statements as well. The event of 2 astrologers making contradictory statements occurred 8 times during that week.

Weasel Words

In reading the many statements made by astrologers, I found many, many uses of "weasel words" preceding predictions. Some of these include "is likely", "can have", "could prove to be", "may have", "possibility of", "may be", "may involve", "could help", "might be", "should be", etc. Why don't they use the word "will" instead of all those weasel words?

Tendency of Statements

Many statements tended to be more like proverbial ancient wisdom than predictions.   One syndicate directed its horoscope writer "to provide uplifting, non-materialistic forecasts with no references to television or other competing media". (Weiner, p. 27) Some of these are good advice, and could have an influence on making the world a better place. For example:

From Joyce Jillson

- Use diplomacy, not force, to settle your differences.
- Avoid doing anything that leads to a guilty conscience.
- It's possible to influence others in subtle ways.
- Learn the art of compromising.
- Laughter is the best medicine should life get too heavy.
- Listen to the beat of your own internal drummer.
- Forget trying to impress others; you do better simply being yourself.
- Respect what other people know; you might learn something invaluable.
- Stay open to new learnings and classes.
- Avoid letting little worries loom large in your imagination.
- Developing a deeper understanding of the human condition can make your life more emotionally rewarding.
- Do what you can to help a young child.

From Jeane Dixon

- Handle important details personally rather than delegating them.
- Youngsters thrive on affection.
- Avoid squabbling with adult offspring.
- Avoid revealing too much too soon.
- Do not hide your light under the proverbial bushel.
- Protect your health by learning more about preventive medicine.
- If a desire for power is all that motivates you, you will not get very far.
- Limit your contact with moody higher-ups.
- Be careful not to neglect your loved ones.
- Service to your community will bring you a special honor.

From Sydney Omarr

- Keep options open.
- You learn through process of teaching.
- Give full play to intellectual curiosity.
- Communicate with individual temporarily confined to home, hospital.
- Experiment with "different" mode of transportation.

From Bernice Bede Osol

- The mind is a remarkable mechanism that can perform wonders.
- If you do not manage your resources prudently today, spender's remorse is likely later.
- Someone who has yet to repay you for what was previously borrowed shouldn't be granted another loan today, unless you're prepared to think of it as a gift – or a lost cause.
- Judge others as you wish to be judged.

From Cosmic College

- For those not moving forward, a bit more education might be the best course of action for you now.
- Let sound thinking prevail in all money-making propositions and agreements.
- Look beyond surface appearances for what might be costly hidden factors.

From Sonia McGinnis

- Focusing on charitable causes, helping those who are less fortunate is especially rewarding.

These statements can be applied every day, not just on the day they were printed.

Omarr and Osol have a particular style for their statements. Osol's incorporate more advice based on lessons learned in life that other astrologers do:

- Don't make promises today for expedient purposes. You may take them lightly, but the recipient will expect you to stand by your word.

- You might find it necessary to form some type of partnership today. This is well and good, provided your cohort is positive and not pessimistic.

Omarr often makes unusually specific statements:
- Message becomes crystal clear by tonight.
- You'll encounter persons with these letters, initials in names: B,K,T
- Clarification received before 11 PM
- Intense activity during hours from 6 to 8 PM.

He also states 4 times during the week that "unorthodox" procedures will bring to pass good things. Omarr's *Los Angeles Times* prints a disclaimer with each column.

Caution

Watch out! If you are one of those people who heeds the words of horoscopes, some statements made by the astrologers can be the wrong advice, for example:
- An old unresolved domestic issue might be resurrected by your mate today. If this occurs, the best thing you can do is to change the subject immediately rather than fueling the fire. (Osol)
- Postpone initiating a lawsuit, negotiate! (Dixon)

On Sunday, October 13, for Cancer, Dixon wrote:
- It would be better for all concerned if others follow your suggestions for today's activities.

What if there were more than one Cancer in the crowd?

Debunk

Horoscopes Flunk Test

Volunteers from my astronomy course at Sonoma State University in Spring 1991 polled people to find out how appropriate their daily horoscopes were.

One student collected horoscope columns on the morning they appeared. We used popular columnists Jeane Dixon (Universal Press Syndicate) and Sydney Omarr (*Los Angeles Times* Syndicate). All indications of date and sign were carefully excised. From them we prepared a sheet of 8 choices for each sun sign. They included the 2 that were intended by Dixon and Omarr for that sign that day ("right sign"), and 6 others by the same columnists selected from elsewhere and elsewhen, mostly from previous days ("incorrect"). During the semester, the "correct" horoscopes were shuffled among all 8 positions to smooth out any effect that position might have.

Other students conducted the polling. Mostly they polled fellow students (average age: late 20s), but also some co-workers and the public. Late in the day, each respondent was given the sheet prepared for their astrological sign that day, with this request: "Considering how your day has gone, please tell how appropriate each of these predictions would have been for you this morning." People only needed to think back over the events of that same day, while those events were fresh and vivid, and test the published horoscopes against them. There was no need to recall events long past. And respondents could judge "appropriateness" by their own standard, just as they would judge horoscopes while reading their newspapers each morning.

The 8 horoscopes were in the first column. The other 4 columns contained blank boxes to check, headed "very appropriate", "somewhat appropriate", "somewhat inappropriate", and "very inappropriate". Respondents were asked to mark how they judged each horoscope.

A faculty specialist in polling thought that our method should produce sufficiently suggestive results, though to obtain statistically and sociologically unassailable results would demand resources far beyond ours – including pretesting several alternate polling forms in large samples, and controlling for many factors of age, gender, ethnicity, etc.

Results and Discussion

In all we got back about 700 sheets with 5,591 responses – usually 8 per sheet, though occasionally someone wouldn't finish.

Here are the results:

	very appropriate	somewhat appropriate	somewhat inappropriate	very inappropriate
"incorrect" %	15%	29%	26%	30%
"right sign" %	18%	28%	26%	28%
"incorrect" #	659	1,292	1,139	1,320
"right sign" #	218	327	303	333

Each night, as I filled out my own sign's sheet, I was annoyed when it had only 1 or 2 really pertinent items, and a whole bunch that weren't, since I expected the bulk of responses to be positive. Most other people thought the same: the only column that gleaned dramatically fewer than the other columns was for "very appropriate" horoscopes. So whether you read the horoscope intended for you, or different ones, the least likely result is that the horoscope you read will turn out to be very appropriate. Perhaps horoscopes are written to sound plausible in the morning, and to be forgettable when they fail. Actual experience demonstrates that the bulk of horoscopes are inappropriate, so to take guidance from a morning horoscope is to misguide your day. We all know what problems we spark by operating on false assumptions!

The results for the horoscopes intended for each date are about equal to those published for the other dates. Everybody else's horoscopes are about as appropriate for you today as the one that claims to be just for you, just for today. Therefore, your sign's horoscope for today is no more correct – and therefore no more valid – than any other sign's, or any previous day's. This means there's nothing special about the one labeled for your sign, so there is no validity to the date- or sign-labeling. Your time of birth does not make any significant difference, and neither does anybody else's, so a major premise of horoscopes – that your time of birth makes a significant difference – is wrong.

This supports the general condemnation of horoscopes by Science. Quite a lot of research has taken apart every allegedly scientific claim. The positions of the planets and stars *do not* influence individual humans on Earth. Only by regarding them as mystical lights in the heavens can you fantasize their influence on yourself.

In a situation familiar to scientists who work with large numbers that build up gradually, the greatest excursion from the overall trend (a mere 3%) is in the item with the smallest quantity: the tyranny of small-number statistics.

What You Should Do

If you like horoscopes, read them all, not just your own. The others are just as appropriate as your own, and you'll get 12 times as much fun from reading them all.

What your should *not* do is change what you'd do because of such a prediction. Most horoscopes are inappropriate for the day, so acting on them would misdirect your attention and energies.

What Newspapers Should Do

These tests demonstrated that newspapers that publish daily horoscopes grossly disserve their readers. What they publish is markedly inappropriate, since more than half of all responses to our poll evaluated them as "inappropriate", and the greatest quantity are "very inappropriate". Any newspaper that publishes such a column severely misguides its readers. Any newspaper that claims to serve its readers' best interests should not publish a horoscope column.

Newspaper horoscope columns, if continued at all, should omit all references to date and Science. As we have demonstrated, all the horoscopes of any day are just as appropriate to the readers as the ones intended for their sign that day, so it is grossly misleading to steer readers arbitrarily to one, compared to the others.

Newspapers that insist on keeping their horoscope column – inappropriate predictions, misleading dates and all – should at least use a disclaimer. "The following astrological forecasts should be read for entertainment value only. Such predictions have no reliable basis in scientific fact." is the warning propounded by the Committee for the Scientific Investigation of Claims of the Paranormal.

Debunk
Planets Out of Nowhere

If astrology actually worked, the influences of undiscovered planets would lead astrologers to discover them. But astrologers never make any scientific discoveries.

In late 1991 a new book arrived: _Ephemerides of X and Y: The Discovery of the Tenth and Eleventh Planets of our Solar System; Astronomical Localization on an Astrological Basis_, by Fabio Francesco Berti, translated from the Italian by Maureen Casey. I had never heard of the author or translator.

The photocopied cover letter identified Berti as a 30-year-old astrological scholar in Verona, Italy. On the first page, Berti asked that I preserve the book because it is important: It predicts, from astrological influences, where astronomers can find the 10th and 11th planets, X and Y.

While Berti explicitly claims that these new planets influence everybody, he found them because of their influence on just 2 people, including himself. What a whopper of an extrapolation! Science often extrapolates, sometimes dramatically, as when we take a few hundred faint nearby red dwarf stars to represent the majority type of star in our galaxy.   But it's always necessary to remember how little is actually observed, because improved observations later may not sustain the trend.

Most of the letter's next 4 pages bemoan prejudice against astrology, and proclaim that clear influences guided his finding of the planets' locations – though he "can send you a full analytical demonstration of the mechanisms", meaning that neither the letter nor the book includes them.

And on page 5, Berti concludes "It is, therefore, neither unfair nor indirect, but only extremely modest on my part to fix a price for the discovery of these ephemerid[e]s. This price is equivalent to the mere printing costs of the enclosed booklet containing them and that is 38 dollars to be sent by cheque to my address." $38 seems awfully stiff for an 85-page paperback.   Is the "shareware" concept of pay-if-you-use-it hitting astrology?

Normally I stash astrological input into a file, and let it see the light of the Sun (or planets) only when someone needs to be convinced of how invalid it is.   But this book is different.

Berti claims there are planets in 2 specific locations.   Why not look there and see if they exist? If I look and see them, he's right; if I look and don't see them, then, to the limits of my instrumentation, he's wrong.   That's as straightforward as Science gets. Instead of dismissing this prediction out of hand, it should be easy to dismiss from actual observation.

An infamous historical lesson supports this. The "Neptune Scandal" of 1846 featured 2 superb mathematicians – Urbain J. J. Leverrier and John Couch Adams – who traced gravitational influences on Uranus to the previously unrecognized Neptune.   It also featured the stuffy, authoritarian Astronomer Royal, Sir George Biddell Airy, who dismissed the predictions without bothering to check them. Neptune was discovered by the very first astronomer to actually look where the predictions pointed, using a telescope capable of revealing the planet. Berti probably remembers this episode, because he calls his publisher "Edizioni del Nettuno" – Italian for Neptune.

I have always disliked the stuffiness of Airy, who refused to even look, so, though I saw no reason to believe Berti's position, I thought I would look. If I was wrong and Berti was right, my consolation prize would be to discover 2 new planets ... in which case I'd even send him his $38.

But it wasn't that simple. Berti's notions of planet motions neglect many critical factors. One problem is the way he suggests consecutive annual positions, failing to leap through the inevitable retrograde loops.

A second difficulty is how Berti states (on page 20) that Planet X slowed down about 1984, while Planet Y sped up in 1972. Celestial mechanics can't permit abrupt changes, though elliptical orbits would let them revolve at continually changing rates.

A third problem is that Berti makes no attempt to reckon distance. From Kepler's Laws and the average rates of motion he gives, his Planet X should be roughly 36 times as far from the Sun as Earth is, and his Planet Y about 93 times as far. For comparison, Neptune and Pluto were about 29; Pluto goes as far out as 49.

Berti gives the planets' positions in a form not conventional in astronomy. Deciphering them forced some assumptions. Berti supplies the right ascension, but neither declination or time. In a table of conjunctions and oppositions, he seems to use coordinates that precess with the Earth instead of the sky, or at least coordinates of the current era. He also supplies astrological geocentric longitudes, and with a simple formula converts those into what he calls "right ascension". It is not clear whether he uses the correct plane, since longitude is counted along the ecliptic, while right ascension is counted along the equator, which is inclined 23.4° to the ecliptic.

To determine what positions he really means, I assumed he intends his planets to lie on the ecliptic. For January 1, 1992, he intended his Planet X to be at 7 hours, 16.5 minutes of right ascension, and his Planet Y at 6h 31.8 minutes. As far as I can tell, the declinations should be 22° 16' N, and 23° 13' N, respectively. Both positions are in the modern constellation of Gemini, though in the ancient sign of Cancer.

Of course, precision is probably wasted, since Berti also states in his letter that "When astronomers make [known] the position of the tenth and eleventh planets in our solar system, they will not be dis[s]imilar to those indicated here." How close is "not dissimilar"?

Just going out and looking is also not quite as straightforward as it sounds. The Gemini location poses a problem for observers: The Milky Way's myriad of dim stars compete for attention with a dim planet at Y's location.

I had access to a telescope well-suited to such planet hunting, so I reserved time on it. It had recently been cleaned and optimized. Its 51-cm lens delivers 20 times as much light as the telescope that William Herschel discovered Uranus with in 1781. It gathers almost 5 times the light of the telescope with which Johann Galle discovered Neptune, and more than twice the light of the camera with which Clyde Tombaugh discovered Pluto.

How dim would X and Y be? Not knowing their diameters or reflectivity, that is impossible to compute. If they are similar to Uranus and Neptune, then X should be roughly 9th magnitude, and Y about magnitude 10.5 – within easy grasp of lesser telescopes. If this telescope didn't reveal them, they're awfully puny.

The evening of January 8, 1992, was clear and dark, with little interference from moonlight. The air was quite clean, between storms, so there was little dust or haze for light pollution to reflect off of.   Turbulence (the "angry Jello" effect) was moderate – a bit annoying, but not terrible.

I looked carefully at the positions Berti prescribed. I looked at several fields-of-view around them. I saw plenty of stars which were already mapped. I didn't see anything not on the charts, or that looked the least bit planetary. As far as I can tell, Berti's planets do not exist in the sky.

Berti could claim they're fainter than that scope could find, or that the planets are not exactly where he said they would be.   Because technology is limited, Science cannot state positively that no objects are in the positions Berti states, only that no *detectable* objects are.   No other astronomer has reported any new planets around there, either – though the area is frequently patrolled by big telescopes and satellites. Nor is there any hint of gravitational tug on any known objects.

Was it worth the effort? As astronomy, no. As an exercise, perhaps marginally, though I don't feel the need for it myself and wouldn't assign it to a student. It was to bring up issues worth writing about, and to personally debunk the astrological claim.

Oxymoron
Quarter Moon
by Brad Schaefer

When the side of the Moon that we see is ¼ sunlit, it is a "Crescent", and when it is ½ sunlit it is a "Quarter Moon". The term "Quarter Moon" refers to being ¼ of the way from New Moon through the monthly cycle of phases. But if phases *all* tell the fraction of the monthly cycle, you'd expect that a "Full Moon" has gone through the full cycle, back to "New Moon". Not so. The term "Full Moon" jumps to a different perspective: the fraction of the Moon's disc that Earthlings see sunlit. Thinking of the sunlit fraction of the disc, a "Quarter Moon" phase really illuminates a "Half Moon".

This table shows how much of the Moon an Earthling sees at various times through the monthly cycle of phases. The situations that spark the phase-names are **bold**.

Name of Phase	fraction through monthly cycle	fraction of disc that Earthlings see sunlit
New Moon	0	0
First Quarter	¼	½
Full Moon	½	1
Third Quarter	¾	½
New Moon	1; restart at 0	0

Oxymoron
Waning Crescent

The slender phase of the Moon is beautiful, and the expression *waning crescent* sounds innocuous, doesn't it?

But *crescent* comes from the same Latin root as the musical term *crescendo*: it means "growing". A *crescent* phase therefore is a growing phase.

That makes sense when applied to the sunlit portion of the Moon just after New. To call that a *waxing crescent* is therefore repetitive, redundant and reiteratory, because *waxing* also means "growing".

Unfortunately, astronomy also applies the term *crescent* to the *waning* phases after Third Quarter, when the illuminated Moon steadily thins. You can't tell from the term *waning crescent* whether it's fattening or thinning.

In modern astronomy, *waxing* and *waning* still mean what they say. But *crescent* now means "less-than-half-lit."

Copernicus
according to my least-attentive students

Copernvics theory claimed that the sun was in the center of the earth.

[In Copernicus's theory, planets] become brighter and more visible as the earth move further from the sun a less visible as the earth move closer to the sun.

The moon is the only thing that goes around the sun, this is called the Copernican Theory. His book proved the Ptolemaic Theory wrong and that's what we believe in to nowdays (that the solar system is the center of the universe).

Kapernicisses great achievement was when he discovered another planet by looking up at the sky and seeing a comet.

The Theory is Wrong, But Handy

Soon after Nicholas Copernicus published his great book *De Revolutionibus* in 1543, he died. This prevented the Catholic Inquisition from punishing him for his heresy in moving Earth out of the center, and making it merely one planet among many orbiting the Sun.

Copernicus's Sun-centered system came somewhat closer than anything Ptolemaic to predicting planet positions in the sky. While Copernican predictions were noticeably closer, they were still not exact. We now know the big problem was the shape of the orbits: Copernicus clung to the ancient presumption that orbits must be "perfect" circles. They aren't, but nobody knew that in the 1500s.

Though the Roman Catholic Church emphatically denied Copernican theory – even placing it on its *Index of Prohibited Works* from 1616 to 1835 – they did permit using it as a handy-dandy computing technique for improved results; it simply must not be taught as "true". 'Go ahead and compute that way to get the best results, but don't *believe* the system.'

With 20/20 hindsight, some academics have snickered at this, because we know the Earth is *not* the center of everything. But carry the story a few chapters further:
- Tycho makes the sharpest positional measurements,
- Kepler determines from those that orbits are ellipses, and
- Newton derives Kepler's Laws from his own Law of Universal Gravitation.
- Centuries later, Einstein overthrows Newton, regarding gravity as warps in space-time.

To calculate the path of anything moving many percent of the speed of light requires Einstein's equations; that's how they found out that Newton was wrong. But almost everything that astronomy deals with moves less than 1% of the speed of light. At such slow speeds, the numbers from calculating Einstein's formula are identical with the numbers from calculating Newton's simpler formula. So, even now, practically everybody calculates with Newton's formula, and reserves Einstein's more complicated version for the rare cases where things move really fast. They know Newton is physically wrong, they just use it as a simpler way to compute and get the same result.

What these modern astronomers do is little different from what the Church advocated centuries ago: go ahead and use the handiest formula that gives the best result, but don't believe that it is physically true. To be fair, they should stop snickering at that old Church policy, or start snickering at themselves.

Figuring Out Planetary Motions
according to my least-attentive students

Everything looks like there heading eastward but it's only retrograding.

The rectograde movement is an elusion, because we see the plants from the flat moving object which is earth.

The first concept was the helicentric view that the sun, moon, stars, and planets evolved around the earth. But this could not explain the eclispes the longer cycles during the appearing of the North Star.

Kepler was the best mathmaticia of that error.

Kepler's Laws
according to my least-attentive students

[Kepler's second Law] at any given point, the distance of a planet travelled on it's orbit is the same for the same amount of time.

The orbit of a planet sweeps over areas of the ellipse of orbit in the same amount of time.

[Kepler's] law 2 states that equal areas move at equal times.

2nd law — the line that passes through the orbits passes through equal area.

The area of the ellipses from the Sun to another object is the same.

[Kepler's] second law states that two planets moving around the sun will sweep through equal distances in equal times.

[Kepler's] second theorem states that planets move in an elliptical path that covers the same area regardless of its shape, if the arc length is congruent.

The distance from the sun to the earth sweeps across equal time and space.

The square of the semi-major axis was proportional to the cube of the semi-major axis.

The square of the period of the planet was equal to the average area covered by the planet from the sun.

A student of Arthur Upgren once answered the exam question "State Kepler's Third Law of Planetary Motion" with: "Kepler's third law states that when you heat up a star, it jumps to the next higher orbit."

Oxymoron
Laws of Planetary Motion

Ancient skywatchers noticed that most of the sky's points of light stay in fixed patterns relative to each other, but a few bright ones move among them at varying speeds and in apparently irregular ways, as described in "Wondering About Wandering".

They couldn't predict just when, or how fast, they'd be moving which way or where. They classified these moving lights as unruly and unpredictable, and called them *planets*, from the Greek *planetes*, meaning wanderers.

Then, in the 1600s, Johannes Kepler and Isaac Newton uncovered the regularities that underlie the seeming disorder of planets' motions. This greatly satisfied astronomers. But it yields an oxymoron, the *Laws of Planetary Motion* – in other words, the rules by which the wanderers don't wander.

"According to Newton, How Does Gravity Work?"
according to my least-attentive students

[Gravity] is a pulling type thing.

Gravity is a force that pulls on the planets orbiting around the earth.

The product of its masses is directly proportional to the product^2.

According to Newton the force created between the earth's orbit and the moon's orbit is known as gravity.

The distance of an objects is directly proportionate to its masses when inversely

The speed of a planets orbit is inversely proportional to the planets mass.

Newton ... explained how the masses are proportional to the distance in between them.

Gravity is proportional to the square of the distance of an object to a larger object.

The objects masses are inversely proportional to the distance between them.

The masses of objects are directly proportional to each other.

As the force of an object coming down gravity is an inversely force. ... Gravity is like force pushing down or upward when there is force into it.

The force of gravity is directly proportional to the mass of the object being pulled + inversely proportional to the mass of the object that is doing the pulling.

Force equals the gravity of masses inversely by r^2.

$$F = m^1 = m^2$$

The Gravity of two forces is directly proportional to the sum of their distance and inversely proportional to the square of their distance.

[Newton's] equation ... stated that the direct motion and inverse motion were equal.

Newton explained gravity ... through the question of why the moon orbits the sun.

[Newton] asked himself how come the Sun does not fly away.

Newton says that gravity works by a force ... caused when something is being pulled at one end.

The moon doesn't hit the Earth because there are other bodily masses tugging on it as well.

Gravity worked like when the earth is hit by a cannon it would hit somewhere differently everytime.

The force from the object uses its momentum to be pulled or moved. It's an issue of weight. The object or the planet uses a gravitational pull to create movement.

[Gravity] was a kind of centripetal force that caused objects spinning on axis' (planets) to hold each other in place – but orbiting – continuously. The spinning creates a force that keeps the objects from falling into each other.

The same force that pulls an object up must pull it down.

The sun has the greatest mass so the planets closest to it need have more mass than the one's farthest.

The force of gravity is the product of the masses of 2 objects, divided by the square of their masses.

The force of gravity is directly porportional to the sum of the product of two masses and indirectly porportional to the square of the difference of those two masses.

[Newton] credits [gravity] with keeping the planets and the rest of the celestial bodies in orbit around the moon.

Gravity is the force that makes sure that the planets don't fall or hurt anything. That doesn't mean that the things in the sky or anywhere else can't move it just means that they stay in the sky.

Gravity is ... proportional to the object and proportional to the movement of the object.

If you had a gun and shot it down from a big hill, then with more gun powder you would be able to go around the world.

As things move around the sun it creates a forces that pulls things toward the earth or the planets.

Depending on it's mass product, the force either repels or attracts.

The mass of 2 objects is inversely proportional to the square of their masses.

Newton believed that the same force that held a apple in a tree was the same force that held the moon in it's place.

Gravity works by the force of the mass' constant proportion, which is in turn multiplied by the distance multiplied by the constant proportion squared, then cubed.

[Newton's] law of motion explained motion by how big an object was squared by how far it was away.

Studying the planets and the energy that it given toward sun, in terms of distance and the atmosphere of orbiting, to see which planet pertain the gravitation pull or energy from the sun, to fastest to the slowest. but time is within universe but the universe isn't true time. it years a decade to come about a relation to the observe calculation.

The laws of gravity are pretty simple. When two opposing forces of equal magnitude meet they repel. If two opposing forces of unequal magnitude meet the lesser force gives way.

The moons gravity counters earths gravity and earths gravity keeps it in momentum.

Gravity being the force in which things in motion are divided by distance.

If something is less than the gravity pull it will be either sucked in or out.

Gravity is the force that keeps everything from falling out of the Earth's atmosphere.

Gravity ... keeps the distance between objects.

Gravity being two objects force between each other and the square of their distance.

[Gravity] is a central energy in the middle of the Earth which pulls, this energy is opposite from another force which comes from outside Earths atmosphere. That energy is the energy that would make things float if not for gravity. By having these two energys pull at each other, it makes it possible to live without a dense gravital enviroment.

Getting the Slant on Obliquity

by John Westfall
San Francisco State University

The "obliquity" of the Earth's axis is the angle between the equator and the ecliptic. This angle is slowly decreasing. Many books call it "23½°", which is correct if rounding by half-degrees. But other books change that into a decimal, "23.5°". That's wrong. In 2000, the obliquity was 23.4393°, so, when rounding to tenths of a degree, that should be "23.4°". Many books on physical geography make the same mistake.

How long have books repeated this? The obliquity dropped below 23.45° in 1917.

What Astronomy Doesn't Know

Venus Spins Backwards, Uranus Spins Sideways, and Triton Orbits Backwards

If the nebula that the Solar System condensed from spun in a certain way, the resulting objects should orbit and spin that same way. All the planets orbit the Sun in the same direction; the Sun and 6 of the planets rotate on their axes in that same direction; 6 of the 7 big moons orbit that way.

Venus, however, spins backward, very slowly. Nobody knows how that came to be, but it should be an interesting story.

Uranus spins about as rapidly as the other Gas Giant planets, but it's tilted sideways. So, for part of its orbit one pole points to the Sun; at the opposite side of its orbit the opposite pole points to the Sun; in between times, the equator points to the Sun. This certainly should influence Uranus's weather patterns, but we don't see those clearly enough to know. More importantly, there is no consensus on explaining Uranus's tilt or how it came to be that way.

Triton, the big moon of Neptune, orbits backward compared to the general trend. This is unstable: it can't last for the age of the Solar System, so Triton hasn't always been in that orbit, and will eventually fall into Neptune. Where it had been before, and how it came to its present orbit, are not understood. All we have are interesting speculations, some involving Nereid, Pluto and Charon.

Eclipses and Such
according to my least-attentive students

The Moon is a very powerful object. Despite it's astrological importance, it is a major source of energy for the productivity of our solar system as a hole.

moon phases and sun phases

When the moon is in the suns shadow you have an eclipse!

It's not known why the sun gets fully covered by the sun, even though the sun is much larger.

[The Sun's corona] become visible only during summer solstice.

half the earth has day while the other half has light.

On earth, we have a Sun.

The earth rotated moved around the earth.

We get night from our moon.

During sunrise a planet like the moon will ionize.

Sperling's 8-Second Law: All Total Solar Eclipses Last 8 Seconds

Everyone who sees a total solar eclipse remembers it forever. It overwhelms the senses ... and the soul as well –

the curdling doom of the onrushing umbra, the otherworldly pink prominences, the ethereal pearly corona. And, incredibly soon, totality terminates.

Then it hits you: "That was supposed to last a few minutes – but that couldn't have been true. It only seemed to last 8 seconds!"

This effect frustrated my first 4 eclipses, and most fellow eclipse fanatics assure me they've been bothered by it, too. Yet tape recordings, videos, and the whole edifice of celestial mechanics all claim that it *did* last the full, advertised 2 to 7 minutes – to within a few seconds, that's what *really* happened.

Where did all that precious time get lost?

Eclipse Watching

True eclipse freaks recognize only 2 modes of life: eclipse expeditions, and preparing for them. They'll devote a year or 2 to perfecting equipment: telescope, camera, weird filters and film; sandproofed, soundproofed, rainproofed (heaven forbid!), and bug resistant. No matter what their expedition sees or does along the way, they'll fret about totality. Will the clouds part? Will * the * equipment * work? WILL * WE * SEE * IT?

The partial eclipse is a tantalizing, exasperating hour and a half. Then the diamond ring forms, gleams and vanishes – and at last they have totality. They gape in awe for just a second, then dive desperately into the sequence, many times rehearsed, of exposures, adjustments, notations so hurried they can only be unraveled from the tape recordings afterwards.

Inevitably, totality terminates too soon, often even before the planned sequence does, and they never make it to their own hard-won free-looking phase. "But I got it on film!" they proclaim, "And I can frame that and glow at it forever –

even though ... I only saw it ... through the ... camera's finder."

The novice and the non-astrophotographer take the hang-loose approach. Restless in the partial phase, they get impatient and even quarrelsome around the 1-hour mark. But in the last 10 minutes they can feel it: totality's a-comin'. The world is darker, oranger; shadows look oddly sharp-edged. There's a nip in the air, the birds are atwitter, and shadow bands go skittering around. The ominous umbra sweeps in, the corona unfolds, the diamond glitters and is extinguished, and "OH * MY * GOD * THAT'S * THE * MOST * BEAUTIFUL * THING * I'VE * EVER * SEEN!" They stare transfixed, all their senses open, trying to take in as much as they can.

Unwilling to concede that totality can't linger past third contact, they keep staring at the emerging solar sliver long after it gets painfully bright. Finally, they must be ordered to look away. Then, limp, with self-satisfied grins, they applaud, or yelp, or shuffle aimlessly and ask where the next one's gonna be and how to get there.

Both styles of eclipse-watching yield the viewer a solid 8 seconds of memory. I replayed all my mental images of my first 4 totalities in about half a minute. And that was after seeing 12½ minutes of totalities. The other 12 minutes just weren't there! Poof!

Transfixed –

The culprit is attention span. If you stare transfixed, your mind, knowing the scene isn't changing, says "I already know that", and refuses to store away the same image yet again.

So the solution is not to stare.

What? Not look at that most marvelous miracle you've traveled umpteen thousand kilometers to see?

No, I didn't say not to *look,* I said not to *stare.*

Pre-record a cassette, timed to start at the first diamond ring. On it, tell yourself what to notice during different parts of your precious few minutes in the Moon's shadow. Notice how the umbra envelopes you, enjoy the diamond ring, then examine the prominences (they're bright, so you don't have to be fully dark-adapted). Next, survey the corona – its general shape, and any outstanding features.

Switch away for a few seconds, to check the colors all 360° around the horizon. Since totality is just starting, it'll be darkest in the west, lighter in the east. Now back to the Sun. Your eyes, now partly dark-adapted, are ready for the corona. Which is the very longest streamer, and how far out can you trace it? Where is the innermost dark wedge? Pick out an interesting pattern of filaments and make a mental engraving of it.

OK, back to the horizon. Sweep around again, and notice how much difference a minute or 2 makes. The west is lightening, foretelling totality's end, and the east is dark, where folks down-path are just now getting theirs.

Finally, back to the Sun. Review the best coronal details. Look again at prominences, since there's a whole different crop of them on the third-contact side. Watch for the pink fringe of chromosphere that anticipates – yes, here it comes – the second diamond ring.

How quickly the corona fades! – and now, even the last of it is going – and it's incredible how bright even that tiny wedge of Sun's surface can be!

And now this eclipse, too, is over. But this time you've won. From each separate span of attention during totality you can savor your 8 seconds of mental replay. If you moved your attention enough times, you'll recall many times that 8-second limit. Yes, Sperling's 8-Second Law can be beaten!

3
The
Solar System

The Solar System
according to my least-attentive students

The solar system is where everything is located.

When a planet first forms it is like a big ball of mucus.

The rock and metal is from asteroids or comments hitting the planets.

The protosun and planetesimals were formed which later became our current solar system. The remaining gas from the compression expanded outward and formed the Universe.

Comparative Planetology

In the early 1800s, before stellar spectroscopy, an astronomy textbook described important stars one by one, telling what (little) was known about each. By the late 1800s, astronomers learned from their spectra that stars were variations on a single theme. Ever since, textbooks have covered stars as a category of related objects, and explore their similarities.

Closer to us, the Solar System's objects used to look like a sideshow of freaks. Its members were discovered in different ways at different times. "Planets" go around the Sun. "Moons" go around planets. Both categorizations only deal with objects' motions, and not with their components and processes – or even diameters, since the moons Ganymede and Titan are bigger than the planet Mercury, and Io, Europa, Callisto, Triton, and Earth's Moon are bigger than Pluto. Originally, "comet" meant "fuzzy" and "asteroid" meant "star-like", dealing exclusively with appearance, not motion or components or processes. And "meteor" meant "in the air", with no clue about what it's made of. Meteors change name twice in a few seconds: for billions of years until they hit Earth's atmosphere they had been unseen "meteoroids". If they don't entirely burn up, whatever remains to hit the Earth are forever after called "meteorites".

Most textbooks still cling to these antiquated categories, even though space age research shows that the Solar System's members are all cousins. They started with the same ingredients in the same cold nebula, and underwent related processes.

Their mass, and how hot they got inside, govern which processes each object underwent, making them the way they are today. So my graph plots mass versus how hot they got, arraying characteristics that are all talking about the same things.

We used to think that stars, planets, moons, asteroids, comets, and meteorites were different kinds of celestial objects. But space-age discoveries reveal that they are all cousins. The recipe for them all is:

Take a Nebula, Condense and Stir.

The processes that make each what it is now depend mostly on how much stuff it has, and how hot it got.

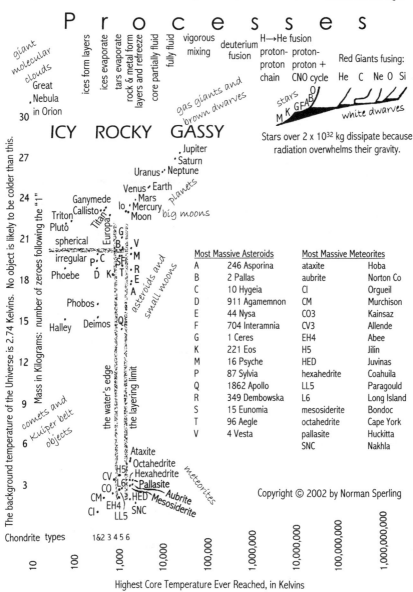

Most Massive Asteroids	
A	246 Asporina
B	2 Pallas
C	10 Hygeia
D	911 Agamemnon
E	44 Nysa
F	704 Interamnia
G	1 Ceres
K	221 Eos
M	16 Psyche
P	87 Sylvia
Q	1862 Apollo
R	349 Dembowska
S	15 Eunomia
T	96 Aegle
V	4 Vesta

Most Massive Meteorites	
ataxite	Hoba
aubrite	Norton Co
CI	Orgueil
CM	Murchison
CO3	Kainsaz
CV3	Allende
EH4	Abee
H5	Jilin
HED	Juvinas
hexahedrite	Coahuila
LL5	Paragould
L6	Long Island
mesosiderite	Bondoc
octahedrite	Cape York
pallasite	Huckitta
SNC	Nakhla

A star core hotter than 8 billion Kelvins collapses in a terminal tantrum, chaotically forging all elements in a supernova explosion, splattering them across the cosmos.

Stars over 2×10^{32} kg dissipate because radiation overwhelms their gravity.

The background temperature of the Universe is 2.74 Kelvins. No object is likely to be colder than this.

Mass in Kilograms: number of zeroes following the "1"

Copyright © 2002 by Norman Sperling

Chondrite types   1&2 3 4 5 6

Highest Core Temperature Ever Reached, in Kelvins

Some of the old categories are distinguishable, and some are not. Comets (retaining original ices) all plot left of "the water's edge". Meteorites are all the small things at the bottom. Stars shine at top right.

But "planets" includes some objects that are physically like brown-dwarf almost-stars, other objects that are like the 7 big moons, and one object scarcely distinguishable from comets. "Asteroids" now have known borderline-cases with comets, meteoroids, and moons; planetologists have long suspected that small moons are captured asteroids and comets, and not original equipment.

Tectonics, subduction, and volcanism only occur on a few differentiated objects. These processes require a rigid (solid, cooled) surface, overlying a warm, fluid interior. On my graph, these conditions occupy a small zone: the smaller planets, and larger moons and asteroids.

- Everything above that zone (more massive) has stayed fluid through the present, so they have no crust on which to show tectonics, subduction, or volcanism.
- Everything below that zone (less massive) is so small it lost heat almost as fast as it gained heat, and probably never melted, differentiated, and formed a solid crust over a liquid mantle.
- Everything left of that zone (colder) never melted and differentiated, so there was no warm fluid to drive tectonics, subduction, or volcanism.
- And no object lies right of that zone because anything that hot is so massive that it plots higher on the graph.

Here is my outline for teaching "Comparative Planetology", covering the same objects and the same data as an object-by-object tour. But since it shows how Solar System objects are *similar* to one another, instead of the old idea to show how they *differ*, it marshals that same evidence in a very different order.

Cosmogony
    Nebular Hypothesis
    Condensation Sequence
    Undifferentiated Material
        Icy: Brownlee Particles; Comets, Chiron, Phoebe, Pluto,
            Charon, D, P, C, & K asteroids and moons
        Rings: Saturn, Jupiter, Uranus, Neptune, Herculina?
        No Ice: Chondrite meteorite types 1-6; S, T, G, B, F, &
            Q asteroids and moons; cometary (shower)
            meteors
        Interplanetary Medium, Zodiacal Light, IRAS cirrus
        Planetesimals
Accretion: Miranda
Differentiation
Differentiated Metals
    Radioactive Core Metals: Earth, Mercury, Venus
    "Dead" Cores: Moon (?), Mars (?), big moons
    Iron: outer cores of planets; M asteroids, iron meteorites
    Core/Mantle Boundaries: M, A, & R asteroids; stony-iron
        meteorites
Differentiated Rocks
    Mantles; olivine meteorites and R & A asteroids
    Crust-Shaping Processes
        Impact Cratering: All rigid surfaces except Io
        Erosion and Soils: Venus, Earth, Mars; "space
            weathering"
        Highlands and Lowlands: Moon, Mercury, Venus,
            Earth, Mars, V & E asteroids; achondrites
        Tectonics: Venus, Earth, Mars, Europa, Enceladus,
            Titania, Ariel, Triton
        Volcanism: Venus, Earth, Moon, Mars, Triton, Io
Volatiles: Surface and Subsurface Ices: Mercury, Earth, Moon, Mars,
    moons of Jupiter, Saturn, Uranus, Neptune
Volatiles: Hydrogen/Helium: Gas Giant Planets: Jupiter, Saturn,
        Uranus, Neptune
    Carbon Dioxide atmospheres: Venus, Mars
    Nitrogen atmospheres: Earth, Titan, Triton
    Sodium atmospheres: Moon, Mercury
Magnetospheres, etc.: Mercury, Venus, Earth, Mars, Jupiter/Io, Saturn,
    Uranus, Neptune

Impacts
according to my least-attentive students

Impacts ... [do] not remain on solar systems.

Impact cratering has taken place on all 23 major bodies in the universe.

[In impact] seismic waves go into the crust at a rate faster than the speed of light.

[Impact can ...] wipe out civilizations such as the dinosaurs.

What Astronomy Doesn't Know
Central Peaks

On the Moon, craters of a certain range of sizes have central peaks. Craters bigger than that size range don't, and neither do smaller ones, and some craters within the size range don't have central peaks, either. Scholars have not reached consensus on explaining how central peaks form, or which craters get them and which don't.

Every other planet and moon that shows craters, has a range of sizes in which central peaks are sometimes found. But for each object, that range of sizes is different.

There must be a reason, but nobody's explained it all yet.

Surface Processes
according to my least-attentive students

The Earth goes through four stages. 1: Flooding. 2: catering. 3: slow surface evolution, and 4: diffrentiation.

Differentiation happens when there is too much dense down and it flows up.

In the middle of the Earth is the crust.

Reset Mindset
Types of Meteorites

Confronted with a variety of minerals that fell from the sky, scientists classified them by what they're made of. All of them are made of stone, or iron, or some of both. So, for hundreds of years, the major classifications have been
- stony
- stony-iron, and
- iron.

This classification is true – there are still no exceptions – but misses the main points learned since the mid-1900s.

Most meteorites are (as long suspected) fragments of asteroids. (A very few are chips of Mars and the Moon. No other source has been confirmed.) There are roughly as many different categories of asteroids as there are of meteorites, but (as of 2002) general agreement on only a few types of meteorites as samples of identified asteroid types.

Though exact identifications still elude Science, the overall lessons become clear when analyzing meteorite contents.

The minerals in most stony meteorites have never melted. They include little rock spheres, a few millimeters across, called "chondrules". These were flash-melted in the solar nebula, before planets condensed. They cooled off very quickly and were cold-pressed into rocks with other minerals. The meteorites containing chondrules are "chondrites". They have always stayed under 1100°C. Chondrites also contain carbon or iron. Chondrites are partly classified by how hot they got. Rare, primitive ones never reached room temperature. Higher and higher temperatures drove off ices, then drove off tars, then almost melted the rock.

The other meteorites have melted and refrozen. When they melted, they differentiated.

Only the term itself, "differentiation", is difficult. The concept is blatantly familiar. In a fluid, dense stuff sinks and light stuff rises. Anything you must "shake well", such as salad dressing, differentiates into layers.

While the planetary object is fluid, the dense stuff – iron and other metals – sink to form a core. The lightest stuff – like gases and water – rises so much that most of it escapes, unless the planet is so massive that its gravity can even hold onto such light stuff. Medium-density stuff, like rocks, form layers in between. Granite and basalt form a crust, while denser olivine forms a mantle.

Millions of years later, cooled off, the object collides with another, and fragments splatter around the Solar System, some of which eventually land on Earth as meteorites.

These meteorites tell about Solar System processes! The chondrites tell about cold, old, primitive objects that have hardly changed since the Solar System began 4.54 billion years ago. All the others tell about bigger objects that got so hot they differentiated. Iron meteorites tell about the cores they used to be parts of. Stony-irons seem mostly to tell about core-mantle boundaries, though a few may be surface melts from collisions. Olivine stony meteorites tell about mantles. Basaltic meteorites tell about crusts.

So the more important categorization of meteorites, and the asteroids they came from, is whether they differentiated or not, and, if so, what layer they were in.

Heavy Metals in Planet Cores

Undifferentiated meteorites contain traces of "heavy metals", the rare elements at the bottom of the Periodic Table that are often radioactive. Differentiated meteorites contain about 1/1000 as much of those heavy metals. Where do they go in differentiation?

It's not a simple problem. Some may actually float up because their ions are large. Many may dissolve in assorted minerals (though no such mineral has been found). Simply by being so dense, they probably sink to the very center.

Perhaps no heavy-metal meteorite has been found because that would require shattering the tough iron outer core wrapped around it, and collisions in the asteroid belt haven't been strong enough. Perhaps they have been, but the heavy-metal inner core is so tiny that we just haven't found one yet.

The core of a large object, like Earth, rich with radioactive metals, heats up from all that radioactive decay. That's why Earth's core is so hot.

The Difference Between an Asteroid and a Meteoroid

... is whether we can see it while it is still out in space.

Asteroids

Any small object we *can* see orbiting the Sun is an asteroid. "Asteroid" means "star-like" in appearance: a tiny dot, showing no shape or features. When first discovered, each is bright enough to register, but too small to show shape.

Until the 1920s many astronomers thought stars were made of the same things as planets like Earth. The suspicion (and, at the turn of the millennium, proof) that asteroids are made of rock and metal similar to Earth has not changed, so asteroids were considered "star-like" in composition as well as appearance for over a century. The comparison to stars' composition was left behind when stars were understood to be glowing gas globes, mostly hydrogen and helium. By then, the "asteroid" name was well-entrenched (though some call them "minor planets" and a few call them "planetoids") and the term hasn't changed even though it is no longer valid.

Meteoroids

"Meteors" are the bright points of light streaking through the air; the term "meteor" means "in the air". If we couldn't detect the object before it zipped through Earth's atmosphere as a meteor, it had been a "meteoroid". After burning through the air, anything that remains to land and be picked up is called a "meteorite". So the same object carries different names, distinguished by when it runs into Earth's atmosphere. For billions of years it had been a meteoroid; for a few brief seconds it is a meteor, and forever after it is a meteorite.

"Meteoroids" are a category which, *by definition,* are never seen. If they *are* seen, they are automatically called asteroids.

The boundary between "asteroids" and "meteoroids" depends on detectability, not on the actual objects. It depends on the telescopes that humans use to look at them with. In 1810, the smallest noticeable asteroid was roughly 200 km diameter. By 1900, 30 km asteroids could be seen. By 1950, 2 km asteroids were found. At the beginning of the 2000s, the smallest asteroid charted was about 7 meters across. In nearly 200 years, tens of thousands of objects changed from undetectable "meteoroids" to detectable "asteroids", though the *objects* did not change, only our *telescopes* did.

Venus
according to my least-attentive students

Venus sits closer the the earth ozone than any other planet, which contributes to the heat.

Venus has few craters and almost no small craters because its 1thousandth the density of the moon.

Venus is a red giant.

Right Physics, Wrong Planet
The Hadley Vortex

When you look at an ultraviolet picture of Venus's clouds, you see a gentle Hadley Vortex: air is heated most along the equator, then curls around to cool off at the poles. This is named for the English scientist who thought it up, George Hadley.

The problem with the Hadley Vortex is that Hadley was wrong. Hadley proposed this pattern way back in 1735, to explain the air circulation of Planet *Earth.* A little while later an American scientist you've heard of, Ben Franklin, demonstrated that Earth's weather systems are much smaller than that. So the Hadley Vortex was discarded, and stayed discarded for 2 centuries. Then spacecraft discovered that Venus's air does circulate like that. So Hadley was right, even though Hadley was wrong: he had the right physics, he just had the wrong planet.

Brad Schaefer continues: Hadley cells don't work on Earth because they're overwhelmed by 2 other effects.
- Earth spins rapidly, which results in "Coriolis" forces that divert the simple Hadley flow into circular motions. Those break up into the spinning low-pressure and high-pressure weather patterns you see on weather maps. This is also why you never see Hadley flow on the Gas Giant planets: they spin so fast that these Coriolis forces overwhelm the Hadley flow.
- The second reason (which only works on Earth) is that there are hot spots (cities, deserts), cold spots (lakes, oceans, icecaps), and wind diversions (mountain chains) that all break up the simple Hadley flow. Venus has none of these (its mountains are small compared to its atmosphere) so no other effect overwhelms the simple Hadley flow.

Planet Earth
according to my least-attentive students

Earth plates collide you either have seduction or a raising of material.

The earth has some primary advantages: a nitrogen atmosphere, relatively stable cluster and liquor water.

Oceans are made up of Platonic plates.

There are several diffrent kinds of volcanoes, sheild, cinder block, & ridge volcanoes.

a cylandar cone volcano

Cinder volcanos have lava that when [water] hits hot rock it changes into steam and creates many cinder blocks.

Earth's magnetosphere is against the gravity from which objects are pushing against us from outside.

4.5 million yrs. ago was when our solar system burst into the atmosphere.

The big bang is what happiens when all life ends on Earth . .... The people on earth have never witness a big bang, but Astronomers have come up with this idea by see great evidence of big bangs happening on Venus.

Mars
according to my least-attentive students

The Big Bang occurred and Mars spontaneously appeared.

Mars is located halfway between the Earth and the Sun.

Mars consists of twelve different rocks. ... Standing aside the highland lowland terrain is the Western Hemisphere of mars. ... Through research Mars has been found to withhold all the necessities required for human development and inhibition. ... Flowing through the habitat would be an artificial river with water recalculating through an aqueduct system parallel to the canyon.

What Astronomy Doesn't Know
Mars's Spiral Polar Cap and Flashes

Mars's north polar ice cap is patchy in a very beautiful and unexpected way: the grooves form a *spiral* pattern. A suggestion of similar arcs can be found in the south polar ice cap, too. This pattern remains unexplained.

Observers occasionally report flashes from certain parts of Mars, and these have even been videotaped. No explanation offered as of 2002 satisfies known conditions.

Gas Giants
according to my least-attentive students

Gas giants suffer from equatorial bulge.

Oxymoron
Frozen Gases

The term (confusing): Since about the 1950s, books say the outer Solar System contains "frozen gases." But physics and chemistry courses teach that solids, liquids, and gases are DIFFERENT states of matter. So "frozen gas" is an oxymoron. If it's a gas, it isn't frozen; if it's frozen, it's a solid, not a gas.

The concept (easy): The outer Solar System abounds in solid water, carbon dioxide, methane, and ammonia. The compounds are very simple. They almost always come jumbled up, sometimes patchy, sometimes smoothly mixed.

Solid water, everyone agrees, is called "ice". Solid carbon dioxide, everyone agrees, is called "dry ice". I don't know any common name for the solids of methane and ammonia. Collectively the solids of water, carbon dioxide, methane, and ammonia can be understood as "ices."

What Astronomy Doesn't Know
How Old is Jupiter's Great Red Spot?

The most prominent storm on Jupiter is the Great Red Spot. Some books call it a 300-year-old storm. That is baloney. 300 years is not the age of the storm. 300 years is how long humans have had telescopes good enough to see it. Now, I suppose, philosophically, you could claim the spot sprang up the day before the first telescope that was good enough to see it. But I doubt that, and so do you. It could be a three-hundred-year-old storm, but it could be a three-thousand-year-old storm ... or even a three-million-year-old storm ... or even a three-billion-year-old storm. Nobody knows! All we know is that it has been there as long as we've been able to see it.

What Astronomy Doesn't Know
The Gas Giants' Stripes

Back in the 1970s, one of the goals of planet probes was to learn why the stripes of the Gas Giant planets look the way they do. The colors obviously result from different chemicals and perhaps different conditions; space probes would tell us *which* chemicals and *which* conditions. The stripes darken and fade, thicken and thin, and sometimes sprout diagonal markings. Those are weather patterns, and watching them up close would help us explain them, and help us understand our own planet's weather.

The space probes worked wonderfully well. The imagery and other data are spectacular.

But we still don't know which chemicals and which conditions make which colors in the clouds of Jupiter or Saturn. And every change in the Gas Giants' cloud patterns still takes us entirely by surprise: that Neptune had a Great Dark Spot; that it disappeared; that assorted belts, zones, and spots come and go on each planet. We still don't understand weather well enough on any planet, including Earth.

Another feature of Jupiter's stripes is also unknown. Many textbooks imply that the stripes – dark "belts" and light "zones" – cover the whole planet. They are indeed very prominent on the middle half of Jupiter, but the closer you look to the poles, the more muted the stripes look. Almost no striping persists to the poles. Nobody knows why the stripes are so soft near the poles. For that matter, nobody knows why they're so bold near the equator.

Io
according to my least-attentive students

[On Io] below the liquid sulfur and SO2 there is a molten sulfur underneath.

Right Physics, Wrong Moon
Volcano Seen On Edge of Moon

I vividly remember when I first saw this famous *Voyager 1* picture of Io. I was an editor at *Sky & Telescope* magazine, and we were oohing and aahing over the latest *Voyager* pictures expressed from the Jet Propulsion Laboratory. When we got to this one, I told the others that I had seen that picture a few months before. They said that was impossible, it had just been taken a couple days before, and no one knew till then that Io had volcanoes.

I went home, rummaged around, and found a classic old book I had bought a few months before: *The Moon,* by James Nasmyth and James Carpenter. I had long wanted the volume, saved up, and bought the next one that came on the market. Having spent so much on it, naturally I sat down and read it. The century-old approach was quaint, to say the least, and the woodcuts and engravings primitive to space-age eyes. How much astronomy had progressed in 100 years!

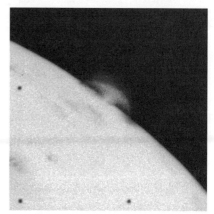

There was the picture I remembered. It depicted a quaint, rejected idea. In 1874, the only way geologists knew to make craters was volcanically. Since telescopes revealed so many craters on the Moon, they concluded that the Moon must be heavily volcanic, at least in the past, because that was the only way to get so many craters. They had no notion of impact cratering.

Nasmyth was not only an expert lunar observer, he was also an artist. To illustrate the way the Moon must have gotten its craters, he drew this picture of a volcano spouting off. The perspective was undoubtedly picked for its artistic appeal.

As decades went by, selenologists and other planetologists realized that the Moon is only minimally volcanic. Almost all craters result from impact. So they rejected the old ideas, and illustrations like Nasmyth's became old-fashioned curiosities.

That status abruptly changed with the Io flyby. Io is volcanic, and we saw the plumes from volcanoes seen on the edge of *that* moon. Nasmyth's picture could hardly have been more prophetic. He had the right physics after all, he just had the wrong moon.

Comets
according to my least-attentive students

Comets ... are in some way if I may say so a way of releaving pressure.

The comet crosses the orbit of a planet, which allows astronomers to document their findings.

Comets can fortell the future such as when they land, in what month or time.

Comets are important because the gases they release are good for the solar system.

Dead comets become planets.

Comets are largely composed of hydrogen. Comets originate from explosions in space or stars combined with meteors falling out of their galaxy. The comet will last longer if it is composed of some iron as well as hydrogen. It's life will last longer as it crosses through the atmosphere if it is larger of course. Comets are capable in finding homes in certain galaxies that are compatible to their existence.

The comits are known from there agressiveness and conviction that make them part of our magnitic fuild process.

Comets always move away from the sun.

Most of the Sun was created by comets that had broken up.

Comets rotate on the outside of the earth's atmosphere but never enters.

Though [a comet's] tail is so massive, it has very little mass.

The tail ... always pointed away from the tail.

Comets were thought to be the pieces which did not adhere to anything and thus fell burning through the sky due to the extremely hot temperatures endured in it's formatting process. Meteroids were pieces somewhat of a much cooler temperature than comets – due to it's lack of flammable material make-up – which were lost in one planets' – or the universes' – gravitational pull then it was engaged into another planets g. p. with greater force.

Debunk
Putting *Worlds in Collision* in its Place

In 1950, Immanuel Velikovsky published his famous book, *Worlds in Collision.* He attributed various biblical events to astronomical phenomena, describing Jupiter, Venus and comets in ways contrary to what Science has learned, and contrary to what was already understood in 1950.

Astronomers immediately denounced the book as unscientific; some urged boycotting its publisher. There was a big ruckus, and many scientists who were unaccustomed to persuasion in public affairs came away bruised by the experience.

Scientists who have read Velikovsky's book tell me it is absolutely not Science.

Even now some people still think scientists suppressed the visionary. They have their own journal, write their own supporting literature, and keep the issue alive, at least in their own minds.

Velikovsky's book is still on library shelves. I certainly don't object to that. Libraries should offer a wider range of books than I personally endorse. But I most certainly do object to which shelf it is on. Both in public and in academic libraries, I find it classified as astronomy, which it is not and never was.

The problem is practical, not just theoretical. Students who base term projects on library books can waste precious weeks developing a paper which I will reject or fail if they swallow Velikovsky's book as truth.

I have asked some public and college librarians about shelving Velikovsky's book in a more appropriate classification. They insist that they are not empowered to do so. Ultimately, book classifications come from the Library of Congress. And while the Library of Congress has occasionally shown wisdom in classifying newer books, an insider once advised me privately that they would not reclassify Velikovsky's because of *political* reasons. Is there really political support for Velikovsky's pseudoscience to be called Science? It's time to set this record straight.

What Astronomy Doesn't Know
The Pingoes of Procellarum

Scholars agree that most of the Moon is bone-dry. Minerals brought back by *Apollo* and *Luna* missions show no signs of present or past water. Some ice has been detected in craters near the poles, where sunlight never reaches.

There are some very odd formations on the Moon. The Marius Hills, in Oceanus Procellarum between the craters Aristarchus and Reiner, don't resemble other hills on the Moon. But they do look like a rare form of hill on Earth: the pingo. In and near the hills lie some valleys that have large heads, and eventually peter out.

On Earth, pingoes are hills of dirt pushed up by the expansion of freezing ice on or just below the surface. On Mars, "head-and-tail valleys" closely resembling the ones near the Marius Hills are said to be caused by the sudden melting of ice just below the surface.

Why not on the Moon? Only because the Moon is dry. But it may not take long to make pingoes or head-and-tail valleys. One slug of water from an impacting comet, melting on impact, freezing just below the surface in the cold of a lunar night, and remelting in the warmth of the next day could make the features – and then the water would evaporate away, leaving that part of the Moon once more as dry as the rest.

The Outer Planets
according to my least-attentive students

Hershals discovery of Uranus was derived from his observations that concluded that Uranus is orbitting Neptune.

Pluto was so small and so far away and probably would not have been noticed except for the fact that it switched orbits with Uranus.

Reset Mindset
Pluto and Planethood

Since the 1970s, a debate about Pluto's planethood has flared up sporadically. Various authorities opine that Pluto is, or isn't, a planet. The media usually call this a debate about Pluto. It isn't. The debate is really about the word "planet".

"Planet"

The word comes from the ancient Greek *"planetes"*, meaning "wanderers": the 7 bright lights that wandered, compared to the 3,000 stars that all circled in lock-step. The inventory of "planets", and the meaning of the classification itself, changes with new discoveries and understandings. Here is a historical table summarizing the changing statuses of various objects which have been called planets:

What Have Been Called "Planets"

	ancient	1543-1781	1782-1801	1807-1846	1846-1930	1930-	physical
Sun	planet	center	star	star	star	star	star
Mercury	planet	planet	planet	planet	planet	planet	rocky
Venus	planet	planet	planet	planet	planet	planet	rocky
Earth	center	planet	planet	planet	planet	planet	rocky
Moon	planet	moon	moon	moon	moon	moon	rocky
Mars	planet	planet	planet	planet	planet	planet	rocky
asteroids 1-4				planets	asteroids	asteroids	rocky
Jupiter	planet	planet	planet	planet	planet	planet	gaseous
Saturn	planet	planet	planet	planet	planet	planet	gaseous
Uranus			planet	planet	planet	planet	gaseous
Neptune					planet	planet	gaseous
Pluto						planet	icy
total planets	7	6	7	11	8	9	

Copernicus recategorized the 7 ancient planets by recognizing that the Sun is the center of the Solar System, and Earth is simply a planet orbiting the Sun. Since then, the term "planet" has only told how an object *moves*: it orbits around the Sun. The term does *not* tell about its *physical* nature, despite the impression you got in grade school.

William Herschel discovered Uranus in 1781, calling it (for want of a better term) a "comet"; by the following year the world's astronomers had agreed, to their surprise, that it is a planet. When Ceres, Pallas, Juno, and Vesta were discovered between Mars and Jupiter, 1801-7, they were called planets, too – textbooks of the 1830s describe "the 11 planets". When Neptune and more little planets were discovered in the mid-1840s, the little ones got demoted to "minor planets" or "asteroids", reducing the number of "planets" to 8.

Each of the 9 objects currently called "planets" (based solely on how they *move*) can now be associated *physically* with objects listed another way:
- Gas giants are akin to brown dwarves, "wanna-be stars" made of the same gasses as stars, but not containing enough of them to spark the stable fusion that marks true stardom.
- Rocky planets are akin to the 7 big moons, all of which are more massive than Pluto. The term "moon" itself merely means "it orbits around a planet"; the word tells nothing about the objects' physical natures. (Many asteroids have moons, too.) Certain little moons like Phoebe and Nereid are probably captured icy objects, and Phobos and Deimos might be, too. Moons turn out to be largely rocky but most also have ices. The *motion* label "moon" can co-exist with the *physical* label "icy".

"Comet"

The "comet" category earned its label by looking "fuzzy" or "hairy"; if an object *looked* fuzzy, it was a comet. The same Greek root word operates in "coma" and "comb". Nobody knew how comets were physically constructed till a fleet of spacecraft visited Halley's Comet in 1986. The European Space Agency programmed its *Giotto* probe to point at the brightest thing in view because, as everybody "knew" from Fred Whipple's 1950 theory, the nucleus was a "dirty snowball" and therefore must be white. That remained conventional wisdom till the moment the images came back. The brightest thing in view was not the nucleus, it was gas jetting off the nucleus. The nucleus itself was black, not white. Surprise! A comet nucleus is *very* dirty.

(That cleared up a little problem from the spectacular 1858 passage of Comet Donati. Detailed drawings showed a black spot at its nucleus. At the time, they called that the "shadow of the nucleus", thinking the nucleus should be bright. We now understand that the black spot was not the shadow of the nucleus, but the nucleus itself.)

The fuzzy appearance of comets comes from gases and dust liberated when the Sun heats up the approaching nucleus. Now, we pay attention to the long-lived nucleus as well as to the fleeting vapors; to the dark object as well as the white veil; and to the physical components, a hodge-podge of ices, crumbly tarry goop, and rock bits. We now regard the nucleus as an "object", and the gaudy tails are more of a "temporary phenomenon when close to the Sun". So research has outmoded the old cultural description that comets "look fuzzy".

Pluto

Pluto was labeled a "planet" for historical reasons, not physical reasons.

Pluto was discovered in 1930 by Clyde Tombaugh, a wonderful person, at Lowell Observatory. He was searching for a planet beyond Neptune and found this object there. Since he found this object where he was seeking a planet, he called it a planet. Nobody knew Pluto's physical character till the 1970s and 1980s.

The planethood dispute is sparked by realizing that Pluto is so icy that it resembles a comet – indeed, there is no known difference. Its orbit is so elliptical and tilted that some astronomers call it more cometary than planetary. Even its atmosphere is indistinguishable from a comet's coma. If astronomers had known, in 1930, what Pluto is physically like, and what comets are physically like, they would have called Pluto "a heck of a big comet".

"Planet" is just a *motion* label. Pluto could be "a comet-like icy body that is so big it is a planet" or "a planet that is physically like a giant comet nucleus".

Reset Mindset
Apples, Oranges, Rocks, and Clouds

All planetary data tables list planets' diameters. However, they don't measure the same characteristic for each planet.

Everybody agrees that both Venus and Jupiter are planets. Everybody agrees that both have rock and metal inside, surrounded by gas, which includes layers of opaque clouds.

For Venus, "diameter" refers to the top of the *rock;* the gas is called the "atmosphere". That's because Venus is so similar to its better-known sister planet, Earth, where we Earthlings walk on the rock surface, and to Mars and Mercury. The diameter that books quote for Venus doesn't include its gasses.

For Jupiter, "diameter" refers to the top of the *clouds;* the rock and metal are called the "core". When we observe Jupiter through a telescope, we see the cloudtops; Saturn, Uranus and Neptune are measured the same way. If the diameters that books quote for these giant planets used Venus/Earth standards and neglected their gasses, only measuring the rock and metal deep inside, the numbers would be much smaller. Those numbers would also be much less accurate, because we don't actually *observe* those surfaces. The layers are computed by theoreticians.

Everyone may agree on the diameters of Venus and Jupiter, but that doesn't mean they're talking about the same things. You can't always compare such planets using the data tables, including derived properties such as a planet's average density.

What Astronomy Doesn't Know
Planetary Magnetic Fields

Some planets have magnetic fields. In 1919 Joseph Larmor developed the idea of self-exciting dynamos inside the Earth and the Sun to account for their magnetic fields. The "Dynamo Theory" suggested that stirring in the hot interiors of planets causes electric currents that generate the fields, somewhat as a dynamo works. The planet's rotation rate also contributes.

The Dynamo Theory predicted that because Venus's mass and core heat are so similar to Earth's, it should have a magnetic field similar to Earth's. Spacecraft have since shown that Venus has no detectable magnetic field. The prediction is wrong.

The Dynamo Theory predicted that Mars, showing volcanoes – surface evidence of a hot interior – should have a magnetic field, though probably weak. Mars spins about as fast as Earth does. Mars has no detectable magnetic field. The prediction is wrong.

The Dynamo Theory predicted that Mercury, being so small, should not have a magnetic field. Mercury spins very slowly. Mercury does have a magnetic field. The prediction is wrong.

The magnetic fields of the Gas Giant planets are not in proportion to their rotation rates or core temperatures, as far as is known.

And the Dynamo Theory predicts that the Sun would have a strong global magnetic field, when it has none; instead, there are small patchy magnetic fields scattered around the Sun, often associated with sunspot groups.

The Dynamo Theory fails to predict observed conditions. Textbooks should omit it and call planetary magnetic fields "not yet explained".

Debunk
Was the Loch Ness Monster an Aurora?

Astronomical effects influence a lot of fields. But specialists in those studies don't always know enough astronomy to recognize what's really happening. Here's an example on a famous topic that no one would expect to have an astronomical dimension.

The highly-publicized hunt for "Nessie", the Loch Ness Monster, interests scientists and skeptics as well as the "crypto-zoologists" who hope that, in addition to the millions of *small* species that (naturalists assure us) remain to be cataloged, there may also be some unusually *big* ones. Discovering big new animals wouldn't violate anything scientific, and it would definitely be cool.

Nessie's setting is well known. In Scotland there lies a long, narrow, deep lake, Loch Ness, famous for its opaque waters. Sporadic reports from locals and tourists suggest that a large aquatic animal lives there, only rarely surfacing. A few ambiguous photographs and a lot of folklore support

Nessie. The local hotels hope the hype continues to draw even more tourists than the pleasant landscapes and local culture earn on their own. Similar phenomena include "Champ" in Lake Champlain, Vermont, and "Ogopogo" in Okanagan Lake, British Columbia.

Just what the creatures might be, if real, remains to be demonstrated. I often heard plesiosaurs suggested, though these large marine reptiles are thought to have met extinction at the same time as dinosaurs, the end of the Cretaceous period, 65 million years ago. No plesiosaur fossils have been found in any later rocks.

"Remember the coelacanth!", the advocates remind us. These large primitive fish were also thought to be extinct, and now we have specimens of 2 species caught live – one species near the Comoro Islands and South Africa in the Indian Ocean, and the other in Indonesia. But the main reason to suspect a plesiosaur was its similarity to the "surgeon's photo", now admitted to have been a 1930s hoax.

A number of expeditions have sought Nessie, using more or less technological devices, and techniques of varying sophistication and likelihood of success. The one that produced the strangest result – often cited as the best scientific evidence for Nessie – was conducted in the summer of 1972. A sonar transducer (which converts sounds into electrical signals) was submerged 35 feet in the dark waters, connected by a long wire to analytical equipment aboard a boat. The transducer's signal traveled along that wire to amplifying electronics aboard the ship. If Nessie swam by the sonar detector, it would say so, even if Nessie stayed out of sight of the nearby submerged cameras. That is objective and neutral: no large signals means no large object, no Nessie; large signals can mean Nessie is there.

An hour after midnight on August 9, 1972, the sonar produced the peculiar strip-chart recording which is most often cited as showing the Loch Ness Monster. Though published[1], this strip-chart is so different from conventional sonar output that even pro-Nessie studies quote the opinions of authorities, and several of those hedge[2]. Items by Rikki Razdan and Alan Kielar in the *Skeptical Inquirer* have disputed the positioning of the transducer (free-swinging or stationary), the stimulus for looking there and then (a dowser's signal), and the interpretation of the strip-chart. The matter remains controversial.

Despite the decades since then, I remember vividly where I was and what I was doing that week. I was in Springfield, Vermont, at the most famous astronomical convention in America. "Stellafane" is intended for people who make telescopes, but every year thousands who don't grind their own flock there too. I was attending my first Stellafane that very weekend. The sky was clear and dark. The Milky Way shone prominently. But everybody's attention was on something else. Brilliant green aurora – "Northern Lights" – flitted all around the sky. This was the finest display I have ever seen – the longest, the brightest, the most detailed and the fastest flickering, covering the most sky, right down to the south horizon.

In fact, this was one of the strongest auroras in decades, occasioned by one of the strongest solar flare outbursts recorded to that time. The Sun had just spat out a lot of charged particles, and they whipped Earth's magnetic field around, causing quite a lot of havoc. The storm induced electric currents in long wires, with many reports of damaging voltage and amperage variations. There were surges in the Canadian electric power grid; a big transformer exploded; short-wave radio communications were gravely disrupted; and sensitive electronic equipment was subjected to surges and flutters and spikes of current. *Sky & Telescope* magazine covered the event with no less than 5 articles, and J. A. McKinnon compiled a whole monograph on the event.

Much of Europe reported aurora and other electromagnetic phenomena from this solar storm. Loch Ness lies closer to the zone of greatest auroral intensity, the "auroral oval", than most of Europe.

The peculiar sonar reading occurred at just the time of the second-greatest peak of magnetic intensity. But the Loch Ness investigators didn't report the aurora. Most likely it was cloudy there, as it is about 90% of the time. Even had it been clear, their attention would have been focused down toward the waters, and it would be entirely understandable if they didn't notice diffuse phenomena occurring behind them and apparently unrelated to their interests. They did, however, note that "the hair went up on the backs of their necks" – an effect well-known in electrical demonstrations – though they interpreted that as "primitive instincts" that "there was something ominous in the loch that night"[3].

One sensitive electronic instrument, using a long wire, did give a peculiar reading just when an exceptionally strong gust of solar wind swept by Earth, just when hair rose on their necks. The least-strange interpretation is that this sonar recorded the magnetic storm, rather than the Loch Ness Monster. This might explain why the reading from the Loch Ness equipment is so strange that it requires expert interpretations, and why those say different things.

If so, the Loch Ness investigators may deserve a more charitable treatment than some skeptics have given them. They reported what their instrument told them, and that instrument gave a reading that is possible to interpret as data confirming an unusually large object or creature. The hair-raising clue alone was too little to pick up on. The aurora was probably hidden by clouds, and even if visible would not likely attract their attention, let alone their suspicion. And while atmospheric scientists and astronomers would connect the aurora to the strangeness of signals riding long wires, few other scientists would suspect their instruments of telling them anything beside what they're designed to tell.

Absence of evidence is not evidence of absence, so you can still root for Nessie. But the scientific evidence (with the sonar reading resulting from aurora, and the "surgeon's photo" an admitted hoax) is very meager.

Everything people deal with is embedded in a cosmic setting. The better people understand the cosmos, the better they can deal with it.

Notes

1. Scott and Rines, 1975, p 466; Rines et al., 1976, p 31.
2. Rines et al., 1976, pp 36-7.
3. Rines et al., 1976, p 30.

References

- Klein, M., and C. Finkelstein, *Technology Review,* vol. 79, no. 2, 1976, p. 3.
- McKinnon, J[ohn] A[ngus], *August 1972 Solar Activity and Related Geophysical Effects,* Technical Memorandum ERL SEL-22, Space Environment Laboratory, Environmental Research Laboratories, National Oceanic and Atmospheric Administration, Boulder, Colorado, December 1972.
- Razdan, Rikki, and Alan Kielar, "Sonar and Photographic Searches for the Loch Ness Monster: A Reassessment", *Skeptical Inquirer,* vol. 9, no. 2, Winter 1984-5, pp. 147-158.
- —, "Loch Ness Reanalysis: Authors Reply", *Skeptical Inquirer,* vol. 9, no. 4, Summer 1985, pp. 387-9.
- Rines, Robert H., Harold E. Edgerton, Charles W. Wickoff, and Martin Klein, "Search for the Loch Ness Monster", *Technology Review,* vol. 78, no. 5, March-April 1976, pp. 25-40.
- Rines, Robert, et al., "Loch Ness Reanalysis: Rines Responds", *Skeptical Inquirer,* vol. 9, no. 4, Summer 1985, pp. 382-6.
- Scott, Sir Peter, and Robert Rines, "Naming the Loch Ness monster", *Nature,* vol. 258, 11 December 1975, pp. 466-8.

Sky & Telescope magazine articles on this magnetic storm appear in October 1972, pp. 214, 226, and 237; November 1972, p. 333; and February 1973, p. 130.

Conditions Life Likes
according to my least-attentive students

On earth, and throughout our solar system, everything is made up of carbon.

The earth has some primary advantages: a nitrogen atmosphere, relatively stable cluster and liquor water.

Reset Mindset
Habitats for ET

People who ponder where life might get started dwell on this planet's surface, and their conventional wisdom is overly surface-chauvinist. Several gaseous planet environments should also be considered.

Scholars generally agree about life's chemical requirements. Life must be chemically complex. Carbon is the only chemical known to form complex molecules under any conditions. Carbon makes and breaks bonds readily at about the same temperatures at which water is liquid. Carbon compounds useful to the only example of life we know, Earth's, also use oxygen, hydrogen, nitrogen, sulfur, and phosphorus. There may be yet-undiscovered conditions in which other elements can make complex molecules – many people root for silicon – but so far, that's science fiction.

It turns out that the desired conditions and chemicals abound throughout the Solar System and the disc of our galaxy, and, as far as we know, Population I areas of all galaxies.

Venus

Experts agree that life is unlikely on the surface of Venus. The temperature there hangs around 465°C, which is so hot that carbon compounds break down.

However, up in the clouds, the temperature is just right: that's why there are clouds there. Venus's clouds are made of water droplets, and they form up where it's cool enough to condense water. That provides the right temperature, and water provides the hydrogen and oxygen.

Venus's atmosphere is mostly carbon dioxide (providing the carbon), laced with sulfur dioxide from active volcanoes, providing the sulfur. With some of the right chemicals and a comfortable temperature, Venus's clouds are a potential habitat.

Surface-chauvinists may protest that the atmosphere is an "unlikely" habitat, but recall that life on Earth began not on the land but in the warm fluids of the early oceans. Venus's clouds form a warm fluid. Could living forms persist there? Think microbes, or even gossamer jellyfish.

Some protest that the environment there is much too acidic. True, sulfur dioxide and water make sulfuric acid, and there's also some hydrochloric acid and hydrofluoric acid. But I have a gut-feeling that living tissue can survive a strongly acidic environment. And so do you. That's because (as you've noticed) your gut has not digested you. Your stomach has strong acids that break down living tissue ("food"). Since it has not digested the rest of you, the rest of you obviously can survive in the presence of strong acids. (To find out how, study physiology.) If this can happen with us, it can happen in Venus's clouds, too.

Probes to Venus haven't measured the chemistry of the clouds. To seek life there would require a floating probe that could withstand the environment – perhaps an ærogel or ceramic balloon.

Gas Giants

Experts generally dismiss Jupiter, Saturn, Uranus, and Neptune as likely habitats. Their surface temperatures are far too cold for carbon compounds to be active. Attention instead goes to their moons, such as Europa and Titan.

However, deep down inside, each gas giant planet is very hot, reaching thousands or even tens of thousands of degrees. Each of these planets is made mostly of gases. Somewhere between where they're too cold, and where they're too hot, the temperature must be just right!

That's probably not far below the cloudtops. Spectroscopic evidence, and the *Galileo* probe that dipped into Jupiter's clouds in 1995, showed that the right chemicals abound in the gas giants. With the right chemicals, and the right temperature somewhere inside, these, too, should be considered hopeful habitats.

What Astronomy Doesn't Know
Inside Gas Balls

One probe, from the *Galileo* mission of 1995, dipped slightly into the top of Jupiter's massive atmosphere, and found it was hotter and more turbulent than computer models previously suggested.

Those are all the on-site measurements we have for the insides of any gas object in the universe. We observe (but don't entirely understand) neutrinos that come directly from inside the Sun. Everything else understood about the insides of stars, brown dwarves and gas giants – every object in the universe that's much bigger than the Earth, since most of the universe is made of hydrogen and helium – comes from observing surface vibrations and other behavior, and from computer modeling.

Astronomers were clanking out mathematical models of stellar interiors on mechanical calculators as early as the 1940s. As computing equipment improves, more and more factors can be included. And as observations sharpen, reality imposes narrower and narrower limits.

Computer models are still all we have. Some models predict certain characteristics gratifyingly well, such as the lithium abundances actually observed in the smallest red and brown dwarves.

Other models are probably rather farther from reality. That is certainly not the fault of the astronomical theoreticians who crank them out. These are among the sharpest, cleverest humans ever. The problem is that equations for the effects of rotation, mixing, and magnetic fields demand more computing power than early 2000s computers can deliver. So they leave out those factors, or roughly approximate them.

For example, gas turbulence is maddeningly complex. You've seen smoke (tiny solid particles that ride gas currents) rise, break into puffs, curl, billow, and dissipate. Describing mathematically how each particle moves is just too complicated.

Another example is everyday weather – the behavior of the gases we live in. Very sharp scientists have been studying weather for hundreds of years. Yet, in the 1960s, weather predictions for most places weren't accurate much beyond 6 hours, and in 2000, weather predictions for most places weren't accurate much beyond 20 hours.

One blatant result is that most computer models of gas giants and stars show smooth, spherical boundaries between layers. Yet all those objects are hot, and hot objects show internal and surface motion and mixing. Just because computers can't yet account for those factors doesn't mean they aren't there. Such boundaries are probably fuzzy or lumpy or jagged, rather than smooth and sharp. Any time you see cutaways of internal structures, if the layers are smooth, sharp, and spherical, recognize that as an artifact from the computing, not actually characteristics of those objects.

The Meanings of "Metals"

The term comes up a lot, and you have to figure out the right definition from context.

One meaning is the common one: metals like iron, aluminum, gold, and lead – everything below and left of the zigzag diagonal on the Periodic Table of elements.

A second meaning, for describing stars' atmospheres and insides, is "all elements except hydrogen and helium". By the first definition the majority, but not all, heavier elements are "metals"; by the second, they all are. This usage arose when calculating conditions inside stars. Lots of electrons dislodge from all those elements, so in that important way they behave as metals.

The third usage refers to calculating the insides of gas giant planets, not stars. Inside those, electrons dislodge as easily as they do from the metals of the first definition. But they come from compressed hydrogen. That's called "metallic hydrogen".

So inside gas giant planets, hydrogen is metallic; inside stars, "metals" are everything except hydrogen and helium.

4
The
Stars

The Sun
according to my least-attentive students

Sunspots show up on the sun every 11.1 days.

The shape of the sun changes every two days.

We do not understand why [the Sun's] blind spots are on cycles of irregularity that cannot stay constant.

Neutrinos are rises in the surface of the sun and at present thought to be neucleur fields of the sun but there is no study to prove this.

Arthur Upgren quotes a military officer who was directing an observatory: *"You say the Sun is a star, then why can't you see it at night?"*

How Prominences Come and Go

Prominences poise dramatically above the Sun's surface. Their shocking-pink color comes from hot hydrogen. Time-lapse movies show them arising from the photosphere, expanding outward, and sometimes falling back down.

Those movies are carefully selected. Most time-lapse movies of prominences do NOT show them rising up from the photosphere, but rather appearing, out of nothing-previously-seen, in the corona. Some disappear much the same way, others precipitate down to the photosphere, and a few blow outward. It may be more appropriate to think of them as chromosphere-type conditions that appear in the corona, presumably along strong magnetic lines.

What Astronomy Doesn't Know
The Sun

Many aspects of solar science remain unexplained.

- How the corona is heated to millions of degrees, by a photosphere that is only thousands of degrees.
- Why sunspots exist at all.
- Why the sunspot cycle averages 11.1 years – and why it varies.

- Why very long minima, like the Maunder Minimum, occur.
- Why the magnetic polarity of sunspot groups reverses every cycle.
- Why sunspot groups are often structured like elephant herds, with the biggest one in front, the second-biggest at the rear, and the little ones in between.
- Why sunspots appear closer and closer to the equator as a cycle progresses.
- What generates the local magnetic fields across the Sun.
- What makes flares.

Most of these deal with magnetic fields, which are not understood.

Oxymoron
The Solar Constant

Constants are very important in Science. They tell the proportions of things. Something that's constant, by definition, never changes.

Astronomers have been carefully measuring the intensity of sunlight for many decades. They correct for atmospheric absorption and Earth's varying distance from the Sun, and call this intensity *the solar constant*. Astronomers have found that the Sun's output varies slightly. It's not a large variation, or you wouldn't see astronomers unblinkingly describe how the solar constant changes!

Oxymoron
Solar Cosmic Rays

Discovered in the early 1900s, *cosmic rays* were named for the way they hit Earth's upper atmosphere from space.

Back then, these very-high-energy particles were not easy to track back to their presumably cosmic sources.

However, space probes have demonstrated that a lot of cosmic rays actually come from the Sun, rather than from any more distant "cosmic" sources. While admitting the paradox, astronomers still call them *solar cosmic rays*.

Stars
according to my least-attentive students

The clouds and stars had dealing with one another to make the diffrent planets rotate. Doing the light year the discovery of ultroviloet, gamma, radio. etc make the universe move. Egythians was through the ancient goddess and spiritual visible light played a high structured key with forcusing the universe. Tide and celestial make the earth move from the high tide and low tide.

Fusion is what happened when the nebulae and surrounding hydrogenated comets, planets, meteorites and asteroids got to a point that everything was so hot (9.7 kelvens) that the masses of the objects began to grow bigger until they melted (exploded?).

Stars first start out with hydrogen and helium and stay's in space for a while.

Fusion is important in astronomy because it helps with the differentiation of stars. When the carbon and helium fuses together they can create stars. Then it becomes a big chain. Say then a couple of stars fuse together a make dwarfs, after the dwarf comes the giants and so on. Everything is the sky fuses together and that is how the stuff in the sky was created. Let's take another example say sunlight fuses together with an ice cloud then the final product you would get is a rainbow.

The stars are two times as far from us as the sun is.

Red dwarves are gas giants. The bigger ones melt, and go away faster then the smaller ones.

Red dwarf ... are red, orange, and sometimes yellow, this is because they are so unstable and therefore their color changes.

Blue stars are newborn stars and are closer to us, while red stars have longer wavelengths and are farther from us. This is actually the explanation of the Doppler Shift.

Red small dwarf stars get there heat and dencity through its core of energy or mass that contacts hellium, energy, and pressure from the Suns fusion.

Some red giants can be planets such as mars and Saturn.

Red giants like Jupiter and the sun.

We know that stars orbit around planets with it's energy fusion.

The closer the stars are to the sun, the hotter they are and the more massive they are.

When [main sequence] stars are in clusters they have a magnetic pull on each other that eventually makes the stars part and then these stars fuse.

Steller clouds produce Megnetic Fusion with out weights gravity and cause melecular proton to fuse with a hydrogen atom and Hilium wich produce megnetic Fusion.

Through the spectrum the blue has a yellowish orange color and the red has a orangish greenish color.

The sun is are red gaint it. ... The sun or red gaint Is replace by another red gaint before the gaint fussies out and dies.

Debunk

Buying a Star

by Bill McClain, Jim Craig, and Bob Martino
adapted and edited by Norman Sperling from the
"Buying a Star" FAQ (Frequently Asked Question list)
from the sci.astro.amateur newsgroup.

(1) <u>Can I buy or name a star?</u>
<u>Question</u>: Can I buy a star, or have one named?
<u>Answer</u>: No.
<u>Question</u>: But I heard there were organizations that would do this for you. Isn't that true?
<u>Answer</u>:   At least a half-dozen companies or individuals take money to name stars. But they do not own what they

seem to sell. They send you pretty certificates, but those documents have no validity. They are not recognized by anyone else. No government, professional astronomical organization, or international treaty has ever given any company any such authority.

Question: But the company says that these stars are "officially registered" or "copyrighted" with the Library of Congress or the US Patent Office. Doesn't that make them legitimate or official?

Answer: "Officially registered" can simply mean "registered with the star-naming company." No one outside of the company will accept the list of stars and their "names" as valid. "Official" is a word without much legal meaning, so scammers get to use it very loosely.

A copyright can be obtained for almost anything. A lot of printed material is copyrighted each year, not all of it accurate or true – much is fiction. In any event, the Library of Congress and the US Patent Office do not have the authority to name stars, and therefore cannot confer such authority to any private business or person. Some companies strongly imply that they have such authority without actually saying it in so many words. Fancy graphics, claims of a special "vault in Switzerland," celebrity "endorsements" and other techniques create this impression. Read very carefully what they promise (or more importantly, what they do not promise).

Also, the people of the United States makes up less than 5% of the world population. It's arrogant and ethnocentric to think that a private company based in America (or even the US Government) can take upon itself the right to name stars for the rest of the human race.

The International Astronomical Union is the only organization with the right to name anything in the sky. That is part of their official function. They get this right as part of the International Council of Scientific Unions, a function of the United Nations Educational, Scientific and Cultural Organization (UNESCO), recognized by a treaty signed by almost every country on Earth, with American participation ratified by the US Senate. The IAU does not name stars after people.

(2) <u>What do you get for your money?</u>
<u>Question</u>: What do you get for your money?
<u>Answer</u>: Not much.
Essentially, you get a colorful certificate and a sky chart showing a tiny portion of the sky.   For an additional (usually large) fee, you can sometimes buy a copy of a book, self-published by the company, which lists all the names of the people who have given them money.

(3) <u>How can I see the star I named?</u>
<u>Question</u>: I just named a star or received one as a gift and I'd like to see it. How do I find it?
<u>Answer</u>: Seeing the star will be very difficult or impossible.
The stars "named" by these companies are almost never visible to the unaided eye. They can be very hard to find, even with a large, computer-controlled telescope used at a nice dark location.   The celestial coordinates usually included are often inaccurate or not specific enough.   The star charts provided for the customer are sometimes just photocopies from a book, with a black dot circled in red. Often the dot is hand-drawn on the map (making the problem of positional error nearly impossible to overcome). For these reasons it is very unlikely that you will ever see the star you "named".
        Understand   that   no   planetarium,   observatory, university, or astronomer is obligated to show the star to you. They don't get any of the money, after all. If you should find someone willing to try to show it to you, be aware that this person is doing you a big favor.

(4) <u>Will astronomers ever refer to my star by the name I gave it?</u>
<u>Question</u>: Will astronomers ever refer to my star by the name I gave it?
<u>Answer</u>: Never.
The vast majority of stars simply have catalog numbers, and always will.   Astronomers (both professional and amateur) use these numbers because they are easy to look up in databases or catalogs.   There is simply no good reason to name a star so faint it cannot be seen (unless it has very special properties).

The companies that "name" stars do not distribute copies of their books or lists to observatories or universities, so how would an astronomer ever know about the name you gave it? No astronomer will bother to hunt down that Swiss vault. Even if astronomers did get copies of these lists, they would ignore them.

Finally, there is nothing to keep different companies from "naming" the same star after different people. Indeed, one particular star selling company on the world wide web states up front that they sell naming rights to stars without checking to see if another company has already sold them. In this case, which names should anyone use?

(5) How do things in the sky get their names?

Question: You say that I can't name a star, but many things in the sky already have names. How do they get those names?

Answer: The names of astronomical objects are determined by the IAU. Usually, the only time an object is named after a living person is when that person (or people) discover the object (e.g. Comet Levy was discovered by David Levy, who wrote the Foreword to this book; Barnard's Star was discovered by E. E. Barnard, etc.).

Planetary names come from Roman mythology. This also holds true in the case of planetary moons, although many of the moons of Uranus were named after literary characters, mainly from Shakespeare's *A Midsummer Night's Dream*. These names are approved by the IAU.

Star names come to us via historical convention. Most of the stars that have individual names were named thousands of years ago and were cataloged by Ptolemy in ancient Egypt. The names come from folklore, mythology and location (such as Polaris).

Astronomers' catalogs designate stars in several ways. 400 years ago, Bayer gave the brightest stars Greek-letter designations followed by the name of the constellation, such as alpha Centauri, sigma Draconis, etc. Flamsteed catalogued more and fainter stars later in the 1600s, using numbers followed by the constellation name, such as 51 Pegasi, 38 Ursæ Majoris, etc. Later catalogs tallied more and more stars with more and numbers. The IAU coordinates them all.

Craters and other planetary feature names can have various origins. For example, the IAU prefers the names of famous women (particularly in the sciences) to name features that spaceprobes reveal on Venus.

Traditionally, discoverers of numbered minor planets (asteroids) can propose names for their discoveries to the IAU. The IAU rejects inappropriate names. Asteroids named after musicians Frank Zappa, Jerry Garcia and John Lennon were all named by sympathetic discoverers.

Comets are named after the person or people who discover them. Example: Comet Hale-Bopp was discovered at the same time by Alan Hale and Thomas Bopp. There are a few exceptions, as in the case of Halley's Comet (Edmond Halley didn't discover it, he just predicted when it would reappear, which verified Newton's Law of Gravity).

Most objects that were named before the IAU formed in 1919 still retain their names.

For more information on this function of the IAU, see the Royal Greenwich Observatory leaflet *The Naming of Stars* at: http://www.rog.nmm.ac.uk/leaflets/name/name.html

(6) What else can I do?
Question: So what else can I do? I want to (a): do something special and romantic for my significant other, or (b): help myself and/or others deal with the untimely death of a loved one.
Answers: (a): Flowers are romantic. So are chocolates (and they taste better than that silly certificate would). Wine, a fine meal, a stay in a fancy hotel, any of these would do nicely also. Truly, as they say, "it is the thought that counts".
(b): This is difficult. One might suggest donating to an organization that fights the cause of death, like the American Cancer Society or M.A.D.D. Also, many public institutions like observatories, zoos, and museums have fund-raising opportunities where you can make a donation in someone else's name. That person is then honored with a plaque on the wall, or an engraved brick in a walkway.

The advantage here is that your money goes to a good cause. In addition, it's a lot easier for the family to go see their loved one's memorial brick than it is to see one of those extremely faint stars.

If you feel you need to buy something astronomical for yourself or a friend, get a subscription to a magazine like *Astronomy* or *Sky & Telescope,* a book, a planisphere, or tickets to a planetarium show. In this manner, you can connect with the universe of astronomy and get some value for your money.

Other gift ideas:
- Membership in the International Dark-Sky Association
- Membership in a local astronomy club, planetarium, or observatory
- Star atlas
- Astronomy computer programs
- A pair of binoculars

(7) Why do astronomers get so upset by this?

Question: Why do astronomers get so upset by this?

Answer: Many strongly believe that star-naming is fraud or (at the very least) deceptive and morally wrong (depending on the specific star-naming company and the information presented in their advertisements).

No private company or individual has ever been given the authority to name stars.

Question: Who is really hurt by this?

Answer: That depends a lot on what you consider to be harm. The money involved in any given "sale" is generally between $20 and $100. Many believe that as long as the consumer thinks he is really naming a star, he has been deceived and therefore harmed. It is also very difficult to put a dollar amount on the emotional suffering of a person who discovers that the "memorial star" supposedly named for a beloved relative was, in fact, not really named.

Whatever one thinks about the amount of money involved, this practice can cause problems for those who do not share in the profits.

Most observatories and planetaria, for example, get calls or visits from people wanting to see the star they "named". Of course, the institution could refuse to help them and just tell the truth ("Sorry, this certificate is in no way valid. No private company has the authority to name stars."), but what if the star was "named" for a dead child? Suddenly, one is placed in the position of either telling them the truth and breaking their hearts, or going along with their request, showing them a star near the supposed position, not saying anything, and becoming silent partners with the star-namers. Many see this as an ethical dilemma. It can be quite upsetting to the astronomer who has to deal with it.

Sometimes the people who pay to have a star "named" think that astronomers or planetarians are somehow obligated to show them these stars, and become angry if they cannot be found. After all, if the star name is "official," the astronomer should be able to show it to you, right? Then one is placed in a different sort of uncomfortable situation.

Sometimes nothing can be said or done to mollify such a deceived person.

(8) So what are you doing about it?

Question: Why do astronomers allow star-naming to continue?

Answer: That's a fair question. Although many people believe that this practice amounts to fraud, no laws specifically prohibit selling star "names". Therefore, it is very difficult to force the companies to stop. Astronomers are not police officers or prosecutors, and thus do not have the authority to issue "cease and desist" orders.

There are government agencies with the mandate to protect consumers.
- The Federal Trade Commission is one.
- Every state has an Attorney General.
- Many states and cities have Consumer Affairs departments.

These organizations could do something, but they don't seem to regard star-naming as a serious-enough problem to spark action. Only if a lot of people complain will they investigate.

What we *are* doing is informing the public of the truth. As long as everyone knows that you cannot *really* name a star, and that the certificate you receive is just a piece of paper, we will be content. The bottom line is allowing all consumers to make an *informed* choice. Many astronomers and astronomical organizations devote web sites to this issue.

(9) Additional Information
International Astronomical Union Official Statement
The IAU is the *only* organization with the authority to name anything in the sky. This authority was given to them by international treaty. Their statement about the "naming" of stars is clear, to the point, and can be found at:
http://www.iau.org/IAU/FAQ/starnames.html
IPS Official Statement
The International Planetarium Society is a group of planetaria and professional astronomy educators from around the world. They have an excellent *Official Statement on Star Naming* at:
http://sunsite.unc.edu/ips/Starnaming.html
Royal Greenwich Observatory
The Royal Greenwich Observatory leaflet, *The Naming of Stars*, is at:
 http://www.rog.nmm.ac.uk/leaflets/name/name.html
A Star Naming Company is issued a Violation
In May 1998, a star-naming company was issued a violation by the City of New York for "Deceptive Trade Practices". Read about it at the NYC Consumer Affairs Office Web Site!
http://www.ci.nyc.ny.us/html/dca/html/pressstars.html
Star Namers Turn on One Another
In October 1999, a star-naming company filed suit in Illinois Federal Court against another star-naming company. Read the *Boston Globe* article – it's worth a read (especially if you enjoy irony). http://www.ras.ucalgary.ca/~gibson/starnames/globe.html
A Personal Story
Jim Craig, one of the authors of this FAQ, has a personal story about his experience with someone who "named" a star, thinking it was legitimate. A classic example of one of the problems caused by this practice.
http://home.carolina.rr.com/nirgal/buyastar.html

(10) <u>About this FAQ</u>

The information in this FAQ was written by Bill McClain (<u>wmcclain@salamander.com</u>), Jim Craig (<u>jcc@efn.org</u>), and Bob Martino (<u>martino.6@osu.edu</u>). It was updated and edited by Bob Martino. Additional editing for the book *What Your Astronomy Textbook Won't Tell You* by Norman Sperling. Jim Craig is the current official keeper of the FAQ.

The information in this FAQ is presented so that the public will have the knowledge needed to make an informed choice. We have tried to present the facts clearly. Where we have presented our opinions about the facts, it should be clear that this is what we are doing. We've also tried to present a balanced view. The opinions expressed here are not necessarily those of our employers. We claim full protection under the First Amendment of the Constitution of the United States as we speak out about this practice. Read it at: <u>http://lcweb2.loc.gov/const/bor.html</u>

This FAQ is NOT copyrighted! It is public domain. Please distribute it far and wide. All we ask is that if you use it on your web site, you link to the official site named above. If you distribute it in print form, please retain the authorship information as well.

The Dim, the Weak and the Ugly

How does a researcher select what to research? How does an editor select what to publish?

In both processes, the humans involved are often attracted to bright and beautiful objects. For the researcher, "bright" means plenty of light is available, making it practical to take detailed photographs and spectra. For the picture-editor who has to select some items and leave out others, bright and beautiful objects beat dim and ugly ones.

This means that the results reported in textbooks, the press and research journals are not a fair sample.

Red Dwarf Stars

The most abundant type of star seems to be the red dwarf. It's certainly the most abundant type within 25 light years. The very closest star to the Sun, Proxima Centauri, is a red dwarf – but so dim that you need a telescope to see it. Even the brightest red dwarf is too dim to see without binoculars. Since red dwarves are very difficult to recognize, hardly any are known.

For all their abundance, they aren't studied by very many researchers. Compared to other types of stars, they're dimmer, so there is less light to study. They are generally thought to not do much, other than sporadic unpredictable flares, so there is little of interest to attract researchers.

If red dwarves were studied as intently as, say, white dwarves or red giants, would more interesting things be discovered about them?

Thin Nebulæ

Bright, thick nebulæ get lots of attention. For active nests of stars, for beautiful twists and knots, they look great. There are lots of thinner, dimmer nebulæ cataloged, but only a few observers track them down. Mostly, thin, dim nebulæ get ignored.

If thin nebulæ were studied as much as thick ones, would more interesting things be discovered about them?

Dwarf Elliptical Galaxies

In nearby clusters of galaxies, the most abundant galaxy type is the dwarf elliptical. To see even the brightest requires a significant telescope. Beyond 50,000,000 light years, dwarf ellipticals are very difficult to recognize. Because they are small and faint, not many are known.

For all their abundance, they aren't studied by very many researchers. Compared to other types of galaxies, they're dimmer, so there is less light to study. They are generally thought to not do much, having little nebulosity and no big powerful stars, so there is little of interest to attract researchers.

If dwarf ellipticals were studied as intently as, say, spirals or giant ellipticals, would more interesting things be discovered about them?

With Galaxies, as With People, Pictures Show the Most Attractive, Not the Most Typical

People who select illustrations for books, slide sets, and other media naturally tend to pick the most attractive examples. This leads to some important misunderstandings. People looking at the examples tend to think they're typical, when actually they are not.

"Spiral" galaxies, which physically are disc galaxies, are prettiest to most humans. Therefore, the prettiest spirals show up in books and slide sets a lot more than others do. Ragged and less-symmetrical spirals, and elliptical and irregular galaxies, hardly ever get selected, even though ellipticals are very abundant.

Most textbooks include a photo of the beautiful galaxy M 51, the "Whirlpool". This is the galaxy with the most obvious spiral appearance; smaller telescopes (perhaps 35 cm) will reveal its arms than any other galaxy's. Many books call M 51 "a typical spiral galaxy". It is actually one of the *least* typical! Very few disc galaxies have continuous arms that can be traced so far around. Hardly any other bright galaxy has such vivid arms. Enjoy the beautiful view, but don't swallow the claim that it is "typical". It isn't, which is why so many books include it. More typical galaxies don't look as handsome. Editors select the nicest-looking pictures, therefore making the selections anything but "typical".

Barred spirals, too, rarely look like their "typical" case, NGC 1300. That one, again, looks prettier and cleaner than most. That's a good reason to publish its picture, but it's wrong-headed to call it "typical".

Much the same applies to planetary nebulæ, pre-stellar nebulæ, and surface features on planets. Editors (and often researchers) select the brightest and most attractive ones. Dimmer and less-attractive examples may be more typical, but they're less-often studied and shown.

Contest! Open to all!
Identify the "blandest galaxy", "ugliest galaxy", "blandest nebula", "ugliest nebula", "blandest planetary surface feature", "ugliest planetary surface feature", etc. Winners may be published in later editions and on our website, www.everythingintheuniv.com .

The Brightness Careers
of Stars

This new graph shows how stars brighten, and sometimes dim, during their careers.

Depending on their mass, they take different amounts of time to contract enough to spark the fusion that marks stardom. The most massive stars begin shining less than 100,000 years after their nebula begins to contract. The least massive stars need tens of millions of years to contract enough to start shining.

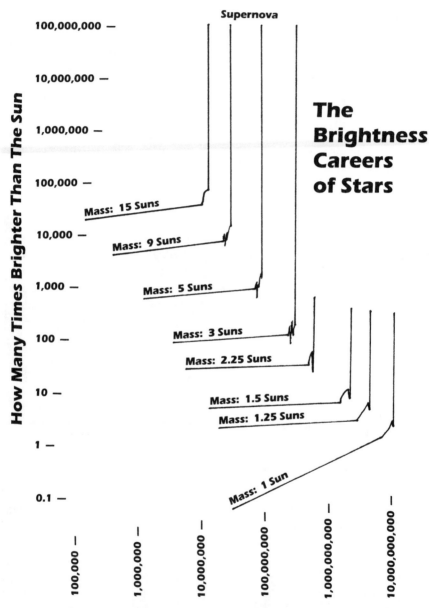

The Brightness Careers of Stars

How Many Times Brighter Than The Sun

100,000,000 —
10,000,000 —
1,000,000 —
100,000 —
10,000 —
1,000 —
100 —
10 —
1 —
0.1 —

Supernova

Mass: 15 Suns
Mass: 9 Suns
Mass: 5 Suns
Mass: 3 Suns
Mass: 2.25 Suns
Mass: 1.5 Suns
Mass: 1.25 Suns
Mass: 1 Sun

100,000 —
1,000,000 —
10,000,000 —
100,000,000 —
1,000,000,000 —
10,000,000,000 —

Years Since Nebula Began Contracting

While on the Main Sequence, stars brighten slowly. In the final few percent of their careers, as Red Giants, they brighten (and occasionally dim) in several spurts – depending, as usual, on their mass. And the most massive stars explode as brilliant supernovæ.

In a cluster where all stars begin condensing at the same time, the most massive stars finish shining before the least massive ones even start.

I assembled this graph from the best data I could find. I will welcome a better graph, or better data to revise this one.

Giants, Dwarves, and White Main Sequence Stars

Tours of the Hertzsprung-Russell Diagram usually declare that "all stars are either dwarves or giants", additionally described by their surface color. Well, almost.

Stars that have evolved off the Main Sequence are generically called "red giants". Many of them really do look red. Some have hotter surfaces, and therefore look orange or yellow, or even white. Whenever someone needs to distinguish among those surface temperatures, they will call those "orange giants", "yellow giants", and "white giants". "Red giants" usually includes all of those, except when context shows they are distinguished by temperature, in which case it only means the ones that actually *are* red.

"Blue giants" occupy the top left corner of the H-R Diagram. They comprise the top of the Main Sequence.

Main Sequence stars from the Sun on down are "dwarves" (or "dwarfs", but that usage comes from English, not astronomy). Those include yellow dwarves, orange dwarves, and red dwarves.

Degenerates below the Main Sequence are called "white dwarves". Most of those really are white, but the hottest ones are blue and the cooler ones yellow. Regardless of actual color, all of them are always called "white dwarves". There is no such thing as a "blue dwarf", and the term "yellow dwarf" refers to Main Sequence stars like the Sun.

Astronomers agree on those distinctions and terms. But notice that they leave a gap. What should we call the white stars on the Main Sequence below the blue giants and above the yellow dwarves? We cannot call them "white dwarves" because that term is reserved for degenerates below the Main Sequence. We cannot call them "white giants" because *that* term means the hottest post-Main-Sequence "red giants". But if we can only call stars "giants" or "dwarves", what can we call these white stars? Surprisingly, there is no generally-agreed-upon term.

As We Learn More About Variable Stars, Textbooks Tell Less

In the early 1900s, variable stars were a hot topic. Textbooks listed many types, interpreted why they behaved those ways, and sometimes gave a taste for the excitement of observing them.

As more and more topics earned their ways into textbooks, "variable stars" often lost page-space. Nowadays, few textbooks spend more than a few pages on this huge topic, nor list more than 3 types. I have seen textbooks leave the impression that there *are* only 3 types: Cepheid, RR Lyræ, and Algol. Those textbooks also cover novæ and supernovæ, which are eruptive variable stars, but don't tell the classification.

Wildly wrong! While pages allotted to variable stars have shrunk, studies of their behavior and nature have enormously expanded.

When discussing white dwarves, novæ and dwarf novæ certainly deserve some space. Flare stars should be mentioned while covering red dwarves, showing there's more happening than we yet understand. And among red giants, Mira-type long-period variables deserve mention.

One type of variable stars has big starspots akin to sunspots (BY Draconis). Another type coughs out clouds of soot (R Coronæ Borealis). And so on.

Variable stars offer non-professionals a way contribute to real Science. With back-yard equipment and competent techniques, amateurs monitor the flickerings of thousands of important variable stars. Professional astronomers use their data all the time.

Multiple Stars Pair Off

Double stars always keep the center of gravity exactly between them. Both zip through the innermost parts of their elliptical orbits at the same time. Both crawl through the outermost parts of their ellipses at the same time.

We know of many triple stars. In every case, they consist of a close pair in rapid orbits, orbited by a far-outlying third member. The third member is so far from the first 2 that it can treat their combined mass as a point. It orbits *another* center of gravity, balancing its own mass against the combined masses of the close pair.

Some quadruple stars are known, most famously epsilon Lyræ. The fourth star always doubles the third: they're always double-doubles.

A few quintuple stars have been found. They always consist of a double-double, orbited at enormous distance by a distant fifth star. That fifth star may be most of a light-year away!

Very few sextuple stars have been found, but every one of them has the same arrangement: the sixth star doubles the fifth. Another way of looking at such a sextuple star is that it is like a triple star, with each member doubled.

Nature always works these orbits as pairings. But nobody has yet figured out how rotating collapsing nebulæ form the multiple-star systems we observe.

Oxymoron
Atomic Fission

Atomic fission may be the most perfect oxymoron ever coined. *Atom* comes from Greek, with *a-* meaning *not*, and *–tom* meaning *cut* or *split* (it's the same *–tom* as in surgical *–ectom*ies and *–ostom*ies.) An "atom" therefore is that which is unsplittable.

When you slam atomic nuclei together, they fuse and release energy. The Sun generates its power by fusing hydrogen nuclei to make helium nuclei.

When you split atomic nuclei apart, however, you can also get energy, though not as much. This *fission* drives nuclear power plants. *Atomic fission* is a wonderful oxymoron: *atom* means you can't split it, but *fission* says you just did.

Where Elements Come From

Elements	Cooked In	Flung Around By
hydrogen, helium, and a little lithium	Big Bang	Big Bang
carbon, neon, oxygen, magnesium, silicon, iron	Red Giants and Supernovæ	stellar winds
more of those plus all others	Supernovæ	Supernovæ

Fusion Confusion: A Burning Problem

Stars spend the first 90% or so of their careers fusing hydrogen to form helium plus energy. Fusion is a *nuclear* process: it only occurs after all electrons have been blasted away from the nucleus, and 2 nuclei can slam into one another. Old stars fuse other heavier nuclei, building up to iron.

Fusion is much more powerful than anything that people normally encounter, so there's no "plain English" term for it. The closest term is "burning", so that's what many lecturers and textbook authors call it.

But "burning" is a chemical process, which chemists call "oxidation". It involves enough energy to get electrons to change the way they're bonded. Burning liberates much less energy than fusion does. It's a different category of process – electron-shell, rather than nuclear. If your mental image of fusion is mere "burning", you're missing most of the energy.

Reset Mindset
Quorbits
By Chris Anderson
College of Southern Idaho

In the early 1900s, the great Danish physicist Niels Bohr proposed that electrons orbit the nuclei of atoms much the way planets orbit the Sun. In the 1920s, however, physicists realized that isn't so. Instead, though it's been illustrated and taught like planet orbits ever since, electrons jump from one place to another according to the probability distribution of their orbitals. It's more of a "probability cloud" that's densest at the places we're most likely to find the electron. Dave Slaven, a high energy physicist, proposes calling these "quorbits" to distinguish a bound electron's relationship to the nucleus from a little planet-like orbit.

Oxymoron
Forbidden Lines

Much of astronomy's evidence for the chemical and physical nature of stars and nebulæ comes from spectroscopy. Astronomers break up starlight into its rainbow spectrum of colors and note which colors are missing (absorption lines) or which are extra-abundant (emission lines).

Each spectral line results from a specific chemical under specific conditions. But scientists on Earth can't produce certain spectral lines in laboratories because they can't replicate all the conditions of outer space. When astronomers found celestial objects displaying spectral lines from elements under extremely unearthly conditions, they called the lines *forbidden*. But the lines aren't forbidden in Nature, they are just forbidden in the laboratory.

The most famous forbidden line puzzled astronomers and chemists for three-quarters of a century.   Many nebulæ show a bright green line at 5007 Ångstroms.   Long suspected to be a different element from any yet known, it was dubbed "nebulium".   It turned out instead to be familiar old oxygen, just more tenuous than any lab can produce.

Oxymoron
The Anomalous Zeeman Effect

Strong magnetic fields split spectral lines.   This fact was first noted by Dutch astrophysicist Pieter Zeeman in 1896. It's called the *Zeeman Effect* in his honor.

After the discovery, Zeeman and other early researchers concentrated on studying the line triplets that classical physics could explain.

But even more astronomical objects have spectra that showed "anomalous" splitting into different numbers and different spacings than classical physics could account for. Quantum mechanics now deals with these effects nicely; they even sparked an understanding of the important concept of electron spin.

Unfortunately for language purists, the earlier term is still used, even though astrophysicists now know that the *anomalous* Zeeman Effect is really more common.

What Astronomy Doesn't Know
Unidentified
Spectral Lines

Physicists and chemists began investigating spectral lines in the early 1800s. Astronomers saw how they reveal the chemicals and conditions of remote objects, and immersed themselves in spectroscopy starting in the 1860s. Ever since, sharp scientists in these and other fields have industriously cataloged the spectral lines of just about all known substances under just about all conditions that laboratories can produce.

At first, some lines appeared in almost everything tested. Scientists would wash and rinse a probe, stick a chemical on it, and heat it. Each substance had its own set of lines, but in addition, they all shared certain lines. Later, they discovered that these were from the element sodium. The water they rinsed the probe with was slightly salty, the salt (sodium chloride) stuck to it when the rinse water evaporated, and the sodium lines showed up along with the intended chemical's.

Using quantum mechanics, physicists have also figured out the spectral lines due to conditions that no lab on Earth can reproduce.

Nevertheless, after 2 centuries of intense research, there are still lines showing up in astronomical spectra that no one can explain.

They can't be from undiscovered elements, a frequent situation in the 1800s, when many holes remained in the Periodic Table of elements. Lines of an element were seen in the Sun's light, and, using the Sun's Greek name "helios", they called it "helium". Years later they discovered helium on Earth. Other unidentified lines from nebulæ were attributed to "nebulium", and from the Sun's corona were called "coronium", but those turned out to be more familiar substances under very unfamiliar conditions.

No more holes remain to be plugged in the Periodic Table, so the still-unidentified lines don't belong to unknown elements. They might be from unfamiliar compounds, or familiar elements in extreme situations.

Oxymoron
Black-Body Radiation

Astrophysicists call an object which totally absorbs all the radiation falling on it a "black body". Physics then tells us that this black body thermally reradiates all the energy it has absorbed. Physicists, with totally straight faces, call this process *black-body radiation.* In other words, in physics every black body emits light, and therefore is not black.

The Pillars
in the Eagle Nebula
according to Dale Cruikshank's
least-attentive students

This exam question was given just a week after a long, thorough description in class – the object's importance, stellar nursery, new solar systems, located near the Solar System in our Galaxy, etc.

This picture shows a region in space photographed with the Hubble Space Telescope. Please answer the following three questions about that picture.
***Number of responses shown in brackets [ ]

A. *Where* is this thing?
a) In the Solar System   [21]

b) Near the Solar System, in our Galaxy [35]
c) In another galaxy far away [11]
d) Between Mars and Jupiter   [6]

B. What is this thing mostly *made of*
a) Iron and other metals [7]
b) Hundreds of old stars [1]
c) Carbonaceous chondrites   [5]
d) Interstellar gas and dust [60]

C. Briefly describe *what is going on* inside this thing

Gaseous materials such as interstellar gas and dust are moving around while keeping an upright pillar-shaped formation, and producing a molecular cloudlike phenomena.

It's bouldery, powdery irregular. Comes from differentiated meteorites. It's either made out of pure iron, iron-silicate, or pure silicate. I don't think much is happening right now since it's still in space. Accretion can't be happening if it hasn't gone through earth's atmosphere.

This picture of the gaseous pillars is constantly breaking off little star-like objects into the solar system. They are chondrules that look like little balls that are melted dust that froze.

Breaking off of this formation of interstellar gas and dust, stellar eggs are forming. Eventually they will break off and explode forming newly born stars.

The picture is about an explosion in Jupiter. An asteroid crashed to the Jupiter and the asteroid exploded. All the dust and gas are spilled up to a high level.

This is and absorption band of various minerals and other materials. It looks like it's glowing on the surface so there are probably radio active elements inside.

The core is being heated to a point where it is melted. The silicate materials float to the top of the melted body. After it cools, it forms meteorites.

This thing is composed of interstellar gas and dust. It is located near our galaxy. The red dots are stars around it. It is like a big blob of energy and this energy inside is what is causing it to glow.

As it moves in its orbit closer to the sun, it will begin to melt. At this point all the metals will differentiate. The silicate minerals will float to the top of the melted parent body. Carbon and complex organic molecules will differentiate to the middle, and the melted iron and nickel will sink to the center.

This old stars burnt up and has turned into interstellar dust which is expanding outward.

If this is what I think it is, a nebula, then there is dust and gases inside. It starts to spin and send bodies into orbit. As the planetesimals and gases spin the collide into each other and attract and influence one another as the motion keeps it moving.

This object is approaching the sun as the glowing cloud around it shows. The nucleus of this object is composed largely of ice which is breaking up as the object moves closer to the sun. The interior is composed of silicate rocks which are similar to those found on Earth which tells us that collisions between objects like this and planets like ours during the first half of the Universe's life is responsible for bringing materials to planets which otherwise never existed.

Inside this large object, the core is slowly starting to heat up inside. Soon it will reach the point where the core is melted and this becomes differentiated. As a result of this occurrence, differentiated meteorites will form. These are created by the silicate minerals that float to the top of the melted parent body.

This is depicting what appears to be an asteroid floating in space. It appears that there is something going on inside like where a comet has ice that evaporates and gives off a glow. This is yellow/orange in coloring and because of some reaction inside the top of it has a glow that the lower half does not have to it. Perhaps this is going to meet the Earths atmosphere soon & burn up or it has hit it. The top may be showing its reflectivity. This is one thing along w/ spectral characteristics that classifies asteroids.

The nucleus of this comet is evaporating because it is approaching the sun. It is losing mass because of this. The gases & ice contained in the nucleus are forming a coma (cloud of evaporating ice & gases) which will form the comet's tail when in contact w/ the solar wind. This is why the object appears to be glowing. In approaching the sun, the coma & tail develop thereby making it a comet.

Nebulæ Contain:

Low Melting Point ("Volatile")

Gases	**H**	**hydrogen**
	He	**helium**
Ices	H_2O	**water**
	CO_2	**carbon dioxide**
	CH_4	**methane**
	NH_3	**ammonia**
Tars	**-C=C-C-**	**hydrocarbons**

High Melting Point ("Refractory")

Dust	SiO_2	**silicates**
	Fe, Ni	**iron, nickel**

Reset Mindset
Nebulous Categories

Classifying phenomena into categories is a standard procedure by which Science seeks to understand Nature's variety. Classifying nebulæ has been a recurring problem throughout the 1800s and 1900s, with more than 25 categories going in and out of use. New categories enter one-by-one as they are recognized. But adopting a new paradigm (usually pushed by technology) may outmode whole sets of them all at once.

Finding Nebulæ

The process begins when somebody notices something. Observers scarcely noticed nebulæ until the mid 1700s. That's because the telescopes used till then had extremely long focal ratios, as much as 300 times their diameter, f/300. They had high magnification, little light-gathering power, and tiny fields of view. Due to imperfect materials and imperfect manufacturing, they also scattered more than a little light. Most nebulæ are so faint and large that the human eye can't notice them when their light is spread out so far that it is diluted that much. Scattered light from stars and planets masked nebulosity. And the fields of view were so tiny that most telescopes only showed a small piece of a nebula at a time, and observers didn't notice that there was a whole nebula there.

In the early 1700s, shorter reflecting telescopes were made, and in the 1750s, John Dollond began selling his affordable achromatic telescopes. Their focal ratios dropped to f/15 or f/20, making a number of nebulæ obvious, and hundreds more detectable as fuzz-blotches. That's why comet-hunting became popular at the same time: comets have similar sizes and surface-brightnesses. Charles Messier, the French observer, could hunt comets so well because he had a telescope with a much faster focal ratio than earlier astronomers. It concentrated light more, and showed a wider field of view, so comets stood out.

This is also why Messier could record nebulæ that no previous astronomer noticed: the telescopes of the late-1700s concentrated nebular light, just as they did with comets. Messier and his successors catalogued over 100 un-moving patches of light, to warn other astronomers not to waste time on them because they weren't comets. Today, hardly anyone remembers the 13 comets Messier discovered. But he is honored for his catalog of things to avoid – the hundred handsomest nebulæ, clusters, and galaxies.

With better telescopes and techniques, the Herschels and others listed thousands more nebulæ. They drew pictures of them, and tried to classify the shapes. Many showed no symmetry. Many were oval. In the 1840s a bigger and better telescope, Lord Rosse's Leviathan of Parsonstown, revealed spiral structure, so "spiral" became a category. The newly-found nebulæ were too faint to take spectra of, so nobody knew what they were made of.

Clouds of Stars

By the late 1700s, improving telescopes resolved some objects (that looked like fuzz-patches to Messier) into clusters of stars. A major question for the next century was whether all nebulæ could be resolved into stars, if telescopes improved enough. In the mid-1800s, spectroscopy proved that some "nebulæ" actually had the spectra of groups of stars, and could eventually be "resolved". However, textbooks kept calling "nebulæ" those which new telescopes resolved into stars clear into the 1880s, decades after forefront researchers relegated them.

Then, in the 1920s, Edwin Hubble proved that spirals, most ellipticals, and a few non-symmetrical "nebulæ" were neither clusters, nor clouds of gas, but whole galaxies akin to our Milky Way, far beyond it. So galaxies were removed from "nebulæ" and established as a separate type of deep-sky object, though textbooks didn't give them a separate category until 1940. (see the Foreword)

Clouds of Gas

On the other hand, many other nebulæ are made of gas, extremely different from collections of stars. The gas ones were labeled "nebulæ properly so called" because not only did they *look* cloudy – the original reason they were called nebulæ – they were thus known to be physically true clouds, a *physical* description.

The ones that looked round and greenish, reminiscent of Uranus, William Herschel (discoverer of Uranus) called "planetary nebulæ", a horrible name we still haven't lived down. That is the only category-name to survive the whole 1800s and 1900s. However, 3 other categories have been merged into it: stellar nebulæ into nebulous stars before 1880, nebulous stars into planetaries around 1910, and ring nebulæ into planetaries in the 1920s.

Making Light

Spectroscopy played a huge role in sorting out these classifications.   So astronomers re-categorized nebulæ according to how they make their light:
- *Reflection* nebulæ simply act as projection screens, reflecting the light of nearby stars. The brightest stars are blue giants, so reflection nebulæ often look blue. The wispy, dusty nebulosity surrounding the Pleiades star cluster demonstrates this type.
- *Emission* nebulæ are heated up by nearby hot stars till they emit different wavelengths of their own. Nebulæ are largely hydrogen, and the resulting hydrogen-pink glow is the most abundant color in the universe. We see this hydrogen-pink from nebulæ and galaxies all across the universe. The Lagoon, Orion, and Swan Nebulæ are famous and beautiful examples.
- *Absorption* or *dark* nebulæ absorb light, and are seen only in silhouette.   (A student once called these

"*omission* nebulæ".) The Horsehead is a famous example; Barnard 86 is often pictured in textbooks, too.

This mid-1900s paradigm still dominated textbooks early in the 2000s, calling nebulæ emission OR reflection OR absorption, as if a whole nebula is one and only one of those. By then, however, many objects were understood to show different parts in 2, or all 3, of these categories, such as the Trifid and Orion Nebulæ. Color photos show emission (pink), and reflection (blue), and absorption (black) areas, all intertwined. In fact, the nebulosity is especially thick in some of the darkest places, such as the lines in the Trifid Nebula, and the "Gulf of Mexico" in the North America Nebula.

- In reality, *all* nebulæ absorb some of the light that hits them.
- *All* nebulæ emit some light (often in the radio and infrared parts of the spectrum).
- And *all* nebulæ reflect some light (though if there's no bright star nearby it won't show up, but that's like blaming a projection screen for not having a projector shining on it at the moment).

So considering the physical manners in which dust and gas handle light, *all* nebulæ are in all 3 categories at all times. Not so useful!

Re-Classify Nebulæ as "Pre-Stellar" and "Post-Stellar"

Meanwhile, stellar astronomers confirmed the old suspicion that stars condense from the nebulæ that show no symmetry. Medium-mass stars die by puffing out symmetrical planetary nebulæ, while high-mass stars die in supernova explosions that leave nebular remnants, too. So some nebulæ turn into stars, and some stars turn into nebulæ. Gasses to gasses, dust to dust.

Nebulæ make sense when taught as part of stellar evolution. First, learn "pre-stellar" nebulæ and how they make stars. Then, learn stellar evolution. Finally, learn "post-stellar" nebulæ such as planetary nebulæ and supernova remnants.

The Pillars With the Fringe On Top

In the Horsehead, Rosette, Eagle, and other nebulæ, black patches show bright fringes. Though clear on pictures taken since the 1920s, no one paid much attention to the bright fringes until the Hubble Space Telescope imaged them in the mid-1990s. The spectacular view of 3 dark pillars in the Eagle Nebula made newspaper front pages all around the world, not because editors understood the picture, but because it is gloriously beautiful.

The pillar tops, and somewhat the sides, show streaky billows of gas coming off the dark pillars. These resemble comet tails. That's because the same process is happening. Both in comet nuclei and in the dark nebular clumps, the dense, black matter is cold, filled with ices, tars, and rock bits. Both in comets and in these nebulæ, nearby hot stars heat up the dark stuff until it vaporizes. Both in comets and the bright fringes, the vapors jet away in wispy streaks. Those wispy fringes resemble comet tails because they result from the same materials under the same conditions, responding in the same way.

Stardeath
according to my least-attentive students

Although white dwarfs are common, red dwarves have only been known within the past million years. Red dwarfs occur only after a white dwarf has died. The red color indicates a new star growth.

White dwarves are not all hot, but they are definitly all very hot.

When iron undergoes nuclear fusion it absorb the energy rather than absorb it.

After the star dies it turns into a meteor.

There are 2 types of supernovas — hot & cold.

Massive stars die because of dust clouds, they eventually break into particles and disintergrate into a black hole. The black hole represents warmth, the climate is warm and allows for a combustile effect for the large floating stars thus forming white dwarfs.

Stars are formed and eventually burn out, whether it is naturally or because it entered our atmosphere.

The star explodes bursting and scattering throught the atmosphere creating a new galaxy.

The stars will die because of having no link with the earth.

Oxymoron
Nova

Over the centuries, skywatchers have reported the sudden appearance of "new" stars, the Latin for which is *nova stella*, usually shortened to *nova*.

New stars? Hardly. Since the mid-1900s, astronomers have understood that these outbursts come from old, dying stars, not new ones.

Novæ occur in well-aged double-star systems, where a degenerate white dwarf draws material from a giant companion star. At long intervals, the transferred material abruptly explodes, brightening the star system several thousand times. The result is called a nova, even though the stars are in their death-throes.

Well then, what about a *supernova*? No-go here, either. A supernova happens when a massive star explodes to end its career. A supernova may be a thousand times brighter than a nova (and hundreds of times more rare), but the object is still nothing new.

What Astronomy Doesn't Know:
Magnetic Fields
by Brad Schaefer

Much of all magnetic phenomena in astronomy is poorly known at best. For example, in gamma-ray bursts, we think we need magnetic fields to explain the (apparent) synchrotron emission, but we don't know the field strength to even a factor of a million, and we have no real idea of how such a field could be created. The same holds for magnetic fields in interstellar and intergalactic space. Also, when magnetic fields have a significant influence on a situation (as with pulsars), the calculation based on the plasma physics in 'high' magnetic fields is too incredibly complex to solve.

A classic question to flummox a colloquium speaker is "Exactly how do magnetic fields change your results?" Most topics harbor *some* effect due to magnetic fields, and this is almost always totally ignored. So this diabolical question will stump almost *any* speaker. It can be asked by anyone, from a wise-and-experienced professor to a beginning student.

Black Holes
according to my least-attentive students

[The dying star] collapses and is condensed in to a single point which is so dense that no light can enter it, which is a Black hole.

Black holes are formed by unknown matters of soilds as well as the gases which contains hydrogen, helium, carbon etc.. Black holes are usually orange stars which defines as old stars.

If you were in a black hole your head would be at your feet.

Black Hole Momentum

In the singularities that make black holes, astrophysicists say that matter loses all identity except mass, charge, and angular momentum. This appears to be true, but incomplete.

"Angular momentum" is the oomph of turning motion. Another kind of momentum is "linear momentum", and though books don't say so, singularities keep that oomph, too. An object that joins a singularity had been moving in a certain direction, so it contributes that momentum to the singularity. If a star is heading in a certain direction when it goes supernova and forms a singularity, the singularity (and the black hole wrapped around it) keep heading in that same direction.

Textbooks would be more complete to say that matter in a singularity only retains "mass, charge and both angular and linear momentum".

What Astronomy Doesn't Know
Inside Black Holes

Inside the event horizon, deep down in the center, the remains of a former star have been crushed unrecognizably. The singularity is so extreme that physics cannot describe the conditions there.

I can't picture a singularity. It is supposed to be a single geometric point. In my mind, I imagine the singularity as the smallest thing I can see – about 1/10 of a millimeter, which is a fraction of the diameter of the period at the end of this sentence. On one hand, I'm wrong: it isn't 1/10 mm wide, it's 0. On the other hand, I'm only off by 1/10 of a millimeter.

When someone figures out what things are like in a singularity, it sure ought to be interesting.

5
Galaxies and Cosmology

The Milky Way
according to my least-attentive students

The Milky Way has a disc surrounding a central bilge.

In the center of the milky way you will find clusters of galaxies. That is surrounded by a halo which is then surrounded by an iodized gas.

The Meanings of the Milky Way

Modern astronomers give the old term "Milky Way" 2 different meanings. You have to figure them out by context.

The classical meaning is "the pale strip of light across the dark night sky". Appearing like a pale white streak, there's no mystery about why it's called "milky"; the name goes back at least to ancient Greek, where "galaktos" (from which we get our modern word "galaxy") meant "milky". This hugs the horizon on Spring evenings, but elsewhen swings high enough to be quite handsome ... if, and only if, you observe from a really dark site. While it is one of the most beautiful sights in the sky, it is one of the more subtle ones, and you really must see it from a dark, remote place to appreciate it. From a city or suburbs, light pollution will glare out the Milky Way, and you'll scarcely see it.

A second, very different meaning has arisen since the 1920s because astronomers learned that the strip of light across the sky results from our location, embedded in a disc of billions of stars. As we look outwards, edgewise into the disc, we see the most numbers of stars, of all brightnesses. The farther away you look from the Milky Way's strip, the fewer and dimmer the stars. This is not an illusion. This is really true. There really are fewer stars to see looking out the thin top or bottom of the disc, than looking along the disc. The disc is the brightest structure of our whole galaxy. Our galaxy's name is "The Milky Way" – the second meaning of the term.

Galaxies
according to my least-attentive students

The [galaxy's] corona is simply crystlized gas.

Three types of galaxies exist, irregular, sperical, and rectangular.

The different types of Galaxies are the Milky Way, the Moon, and with the Sun shining. The types of Galaxies can just about speak for themselves. The Galaxies can be considered as Mars being a Planet.

the diffrent galaxies of the planets

[Disc galaxies] do not contain much mass, but a surprising abundance of matter.

Giant ellipticls are giant things that ellipt.

Our universe are in a disk plane which support each other.

Eliptical Refraction is when there is a Nova behind one of these galaxies.

The galaxies consist of three. elliptical which is the movement spiraling. Our galaxies is within the disk which provides movement against each other.

Reset Mindset
Hubble's Tuning Fork: Out of Tune
By Brad Schaefer

Most texts contain the historic but antiquated "tuning fork" diagram by Edwin Hubble. There is nothing to learn from the diagram, yet it is taught as if it were some Great Truth.

The diagram has no physical relevance. It does not reflect important things like sequences of masses, collisional histories, or angular momenta. It does not represent a sequence (like the Main Sequence) or even steps of galaxy evolution. Real galaxies evolve haphazardly and skip around the diagram frequently as they interact and cannibalize and such.

The diagram also misses many important galaxy types (Giant cD ellipticals, starburst galaxies, Seyfert galaxies, irregulars, dwarves, ring galaxies).

Classifying elliptical galaxies by shape can be very misleading, since a pole-on E8 galaxy would be classed as an E0.

No one ever uses the diagram beyond intro-astro texts, because it has no applications and leads nowhere. Students who obediently memorize the Hubble Tuning Fork should not think it is some major truth by itself.

Reset Mindset
What Shape is M 31?

Messier 31 in Andromeda was first described as "oval" or "elliptical". The reason is obvious to anyone with binoculars or a portable telescope: look at it, and you'll see an oval glow.

Then, in the mid-1840s, Lord Rosse's giant new telescope, the "Leviathan of Parsonstown", gathered more light and saw objects more distinctly than ever before. Enough asymmetric details became apparent to justify reclassifying M 31's shape as "irregular".

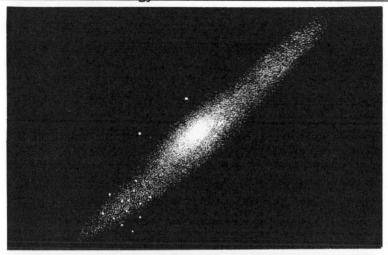

M 31 in Andromeda as "Elliptical"

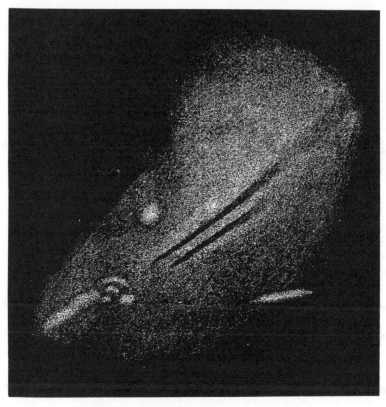

M 31 in Andromeda as "Irregular"

Then, when time-exposure astrophotography revealed the "arms" in the 1880s, M 31 was reclassified yet again, this time as "spiral". Almost every textbook has such a photo.

Textbooks lagged badly, by the way, calling it elliptical as late as the 1890s, some calling it irregular from 1870 to 1912, and generally spiral after 1900.

It was not Messier 31 that changed! It sat there pretty much the same through all those eras. What changed was the instrumentation that humans used to examine it with. The classification depended on the instrument, not the object.

Reset Mindset

What is a Galaxy "Spiral Arm"?

The Milky Way, and many other disc-shaped galaxies, are said to have "spiral arms". The term comes from early drawings and photographs, which show an overall spiral impression in the bright parts.

Humans tend to "connect the dots". When you carefully inspect photos of real galaxies, hardly any have arms so smooth you can actually trace them all the way around.

The very few that do seem to result from recent galactic encounters. M 51 is currently encountering NGC 5195, and M 81 has just passed M 82.

Most other so-called "spirals" only show overall impressions of a spiral-like theme, but notice:
- The "arms" look very patchy. It's easy to notice the bright spots, but make a point of noticing the faint spots

and places where no arm appears at all. Only by connecting the bright patches do people perceive the continuous spiral arms. The actual "arms" are almost always very discontinuous.

- Color photos reveal the bright patches to be blue, meaning they are OB associations. Of course, wherever O and B stars form, every other kind of main sequence star forms too, but the blue giant O and B stars outshine all the others, making the area look blue. O and B stars shine brilliantly but gobble up their fuel much faster than lower-mass types. No blue giant seems to live longer than a few million years. So, what the spiral-arc patches marked by blue giants show is where star-birth happened very recently, and often where it happens right now. Blue giants never live long enough to drift very far away.

- Dimmer stars keep shining long after waves of star-birth sweep over their nests. They fill the disc, including the spaces between the "arms", with lots of stars. While the small patches that trace the "arms" are brighter than the big places between them, lots of light also comes from between arms, especially compared to places way beyond the galaxy. The disc's A, F, G, K, and M stars put out quite a lot of light.

- Color photos reveal arcs of hydrogen-pink nebulæ, paralleling the blue tracery of OB associations. Arcs of dark, thick nebulosity parallel those, too. Radio-telescope traces of molecular clouds also reveal segments of spiral-like arcs.

However, each of those (blue, pink, black, etc.) *lies in a different location!* The disc is full of segments of spiral-like arcs, but no single one of them constitutes a physical arm, because they all lie in different places. *Arms are not physical structures. Arms are illusions. The physical structure is a disc.* On the disc, there are many segments of various sorts, mostly following overall spiral-like arcs.

Some galaxies have spiral segments that appear to be wound much looser or tighter than others. Sometimes the looser-wound segments are nicknamed "spurs". But in some galaxies, like NGC 7217 and NGC 1398, there are large zones where the spiral segments are wound much differently than elsewhere in those same galaxies.

What Astronomy Doesn't Know
A Black Eye for Dust

Many photos show disc galaxies with strikingly bold and stark dark patches on one side, while the other side of the same galaxy looks brighter and much smoother.   This contrast is so strong that M 64 is nick-named the "Black Eye" galaxy. You can find the effect in lots of other galaxies pictured in your textbook. Frequently-pictured galaxies in which you can notice the dark patches include M 31, NGC 7331, and NGC 253.

Even though the effect is very frequently seen, textbooks don't mention it. That's because (as far as I know) no one knows its cause.

What Astronomy Doesn't Know
Why No Giant
Or Dwarf Spirals?

Elliptical galaxies come in dwarf, medium and giant sizes. Irregular galaxies come in dwarf, medium and giant sizes. Spiral (disc) galaxies only come in medium size. Why?

What Astronomy Doesn't Know
How Much are Giant Ellipticals Mergers?

The big brutes of galaxies are Giant Ellipticals. No galaxy cluster has more than 2 of them. They can contain many times the mass of our Milky Way.

How many started out as massive as they have now? How many of them started out smaller? How did they grow? Was it from "mergers" (more vividly nicknamed "galactic cannibalism")?

What Astronomy Doesn't Know
Stellar Populations 3, 1.5, and 0

Walter Baade figured out that galaxies harbor 2 different "populations" of stars. Population I, our kind of place, includes stars of all ages, plus gas and dust. Population II only contains old stars, and almost no gas and dust.

Long later, astronomers figured out that the Population II stars actually came first. They condensed their gas and dust to form stars right away. Some Population I stars started condensing at the same time and more have continued to condense since then, all the way up to now.

Some astronomers, analyzing stellar spectra, have proposed numbering other populations. Population 3 was proposed for extremely old stars made almost entirely of hydrogen and helium, with very little of heavier elements. Population 1.5 was proposed for areas of stars with properties between I and II.

Personally, I wonder:
- If an area where all the gas formed into stars billions of years ago is now a Population II area,
- and if an area where gas is forming into stars right now is a Population I area,
- would an area where gas will form into stars in the future be considered Population 0?

How separate are stellar populations? How much do they blend into one another?

Galaxy Types and Populations

Galaxy type		Star Populations	
Popular Name	**Physical Structure**	**I**	**II**
elliptical	elliptical	(none)	all
spiral	disc	disc	bulge & halo
irregular: *elliptical with oddity*	elliptical	sometimes in the oddity	mostly
disc with oddity	disc	disc, & sometimes in the oddity	bulge & halo
unsymmetrical	irregular	all	(none)

FFNs, LBBs, and LBMs

FFNs

When novices start to use their first telescope, they look at the sky's major showpieces, such as the Messier nebulæ, clusters and galaxies. They're big and bright enough to show up in binoculars, and a beginner's telescope shows detail in many of them. In the background lurk many more faint objects.

Experienced skywatchers buy bigger and better telescopes, seeing ever-richer detail in more and more nebulæ, clusters and galaxies. But always, in the background, there are even more objects, too small and faint to make out. Some irreverent amateur astronomers in San Jose call those background objects "Faint Fuzzy Nothings" – FFNs.

FFNs continue in the background as seen by big, professional telescopes, too. Look at a picture of a galaxy in your textbook. In the background you can often notice dim smudges. Each of those is a galaxy, too, but so much farther away that you can't make out as much detail. A 3-meter-wide telescope shows magnificent detail in objects that amateurs can barely glimpse – and in the background lurk uncountable thousands of more FFNs. A 6-meter telescope shows detail in *those,* and in the background, even more FFNs. A 10-meter telescope reveals detail in *those* objects ... and in the background, there are ever more FFNs. No telescope has ever been made that didn't find more FFNs in the background.

LBBs

One day when I was visiting my brother, a bird-watcher, I noticed his log of sightings. Almost every entry included "LBB". He told me that LBB stands for "little brown bird". They are so common, so small, and so similar, that they're not worth examining to see which common species each one belongs to. They flock all over, they're usually there, and they're not the big or pretty or rare birds that bird-watchers prize.

LBMs

The university's mycological society hosted a meeting about LBMs. Mycologists study fungi, and I didn't have to attend to figure out that "LBM" stands for "little brown mushroom". LBMs are so common, so small, and so similar, that they're not worth examining to see which common species each one belongs to. They're not the big or pretty or rare mushrooms that fungus-hunters prize.

This happens a lot in Science. Beginners learn all the kinds of phenomena in the field, and quickly concentrate on certain ones, all but ignoring certain others. Sometimes practicality forces the distinction: some are available, others are too difficult to study. Often, though, it's about what's *fashionable* to study.

Technology advances at such a furious pace these days that it may be worth looking anew at common background items, using the latest devices. Most people don't pay attention to them. You just might recognize something interesting that no one noticed before.

Oxymoron
The Missing-Mass Problem

One of astronomy's biggest problems hinges on identifying a lot of material that no one can see. Some years ago astronomers discovered that most galaxies, including ours, contain about 10 times more matter than can be seen in any kind of telescope. We know it's there because the gravitation of the invisible stuff tugs measurably on the stuff we can see.

Alternatively, our understanding of gravity might be wrong. But gravitation is one of the most securely pinned down parts of physics. Either that's wrong, or there is stuff we can't see. No astronomer claims we've already seen everything; to say there is yet more to discover doesn't contradict anything.

The discrepancy between what is seen and what is felt is called the *Missing-Mass Problem.*

But the problem is badly named, protests Princeton astrophysicist Jeremiah Ostriker. "The problem is not that the mass is missing. The problem is that it's *there!*"

Lately a few astronomers have started calling this the *unseen-mass problem* or the *missing-light problem,* so there's hope this oxymoron might disappear. Even better, some day they'll identify the stuff and call it by name.

Active Galaxies
according to my least-attentive students

Active galaxies were first observed on the top of Mt. Wilson.

The amount of activity is directly proportional to the galaxies desire to want to be active.

Sephert Galaxies and Sephert Galaxies fall into the same family.

Oxymoron
Radio-Quiet Quasars

In the early 1960s, radio and optical astronomers teamed up to identify sources of radio emission. Until then, the relatively poor resolution of radio telescopes didn't allow them to reckon accurate positions.

Clever techniques identified some pointlike sources that looked like bluish stars, but which had the most bizarre spectra ever seen. Astronomers suspected they aren't stars, so they called them "quasi-stellar radio sources". The term doesn't say what they are; it means "it looks like a star but we think it isn't and we get radio waves from it." Since the term is a mouthful, they used the initials, QSRS. That was quickly pronounced, and then spelled, *quasars*.

Optical astronomers then found lots more of these blue starlike points with similar spectra. However, radio telescopes can't detect most of them. These were promptly dubbed *radio-quiet quasars*. But that term really means "radio-quiet quasi-stellar radio sources". Radio-quiet radio sources!

Cosmology
according to my least-attentive students

The greater the mass of an object the faster it is moving away from the sun.

Our universe was formed by the third star.

The Big Bang Theory ... states that the universe was created due to particles and organisms that lay dormant until they collided, and the Big Bang occurred.

The Big Bang theory states that in order to know what was going on in the universe a million years ago, you would have to have watched it two million years ago.

Nature developed as an explosion in the heavens that fell into the waters and began to grow plants and fish and other underwater creatures.

Before the Big Bang, all the living creatures such as dinasours had been totally dieseased and new birth has been adopted to this new young planets.

There was so much bonding and chemical energy that it all spontaneously combusted and made a universe.

The universe started with that big-bang. A big rock or a galaxy hit the earth and it came to pieces. The fusion up in the galaxy, the pieces, the dust of earth came back together. Before the big-bang, the earth was without water, only dust and volcanos and was extremely hot. After the big-bang, oceans were discovery. The bacteria from the water of oceans transform dinosaurs. The water which have H_2O made the air as oxygen. So we can breath. Soon, the ocean's water wet it the sands, that it started growing plants on the sands and later it became trees and then a forest. The leaves from plants and trees were food for the dinosours. There was a big earthquake that opened up the lands *and swallowed all the dinosaurs. Later the bacteria and germs started to form in molecules and human being started to form. That's how the universe was form.*

When density increases the university begins to contract everywhere.

Every concept is still a theory until it can be proven false.

Oxymoron
The Hubble Constant

Starting about the 1940s, cosmologists have sought to measure how fast galaxies move away from one another. Edwin Hubble pioneered this work, so the number that defines the rate is called the *Hubble constant,* customarily quoted in kilometers per second per million parsecs.

The trouble is that cosmologists still haven't pinned that number down! Hubble himself claimed it was 500, but over the years the constant dropped to 200, then 120, then 70, and a few years ago was claimed to be as low as 47. Constant, it has never been! More than one astronomer has sarcastically called the number the *Hubble variable –* though that's dangerously similar to "Hubble's variable", a pre-main-sequence variable star that Edwin Hubble found in the constellation Monoceros.

Some wise-guy astronomers derived an equation that described the decline of the constant, and called it *Hubble's 2nd-order constant.* Many astronomers have taken to labeling it the *Hubble parameter.*

The battle continues to rage – but, as with the missing-mass problem, a solution may yet emerge. Measurements appear to be converging in the 60s.

Brad Schaefer continues: The Hubble Constant case is even worse. It measures the expansion rate of the Universe, but that keeps changing as the Universe evolves. The mysterious cosmological constant (or whatever the dark energy is) speeds it up, and the gravitational pull of the Universe's mass slows it down. (Many speculative models even suggest that the cosmological 'constant' varies over the age of the Universe.) Generally the term refers to the *current* expansion rate. But even with this reasonable dodge, it changes (very very slightly) over our lifetimes. A better dodge would pin the Hubble Constant to one time and refer all calculations to that.

What Astronomy Doesn't Know
Did the Big Bang Make Mini-Black Holes?

Some versions of the Big Bang theory suggest that the enormous power of the expansion, combined with the extreme density of everything just after expansion started, might have made pockets *im*plode to form black holes less massive than the kind made nowadays by supernova explosions.

Mini-black holes would be even harder to detect than stellar ones. They could be anywhere, but exert extreme influence over such a small zone that scarcely anything we could see would reveal them. In other words, we're clueless. But absence of evidence is not evidence of absence.

A few particle physicists suggest that the "pair creation" process would lead these black holes eventually to evaporate in a flash. The less mass, the faster the process. Given the apparent age of the Universe, black holes the mass of a mountain would be flashing now. No observed flash has yet been proven to come from such a source.

Appendix:
Project Suggestions

Many students don't know what to do for a project. If that's you, consider these suggestions. Though many emphasize cultural aspects, always include and explain the relevant science. Some are tried-and-true favorites. Always make sure your instructor approves in advance of you doing the one you select. Always make sure you know all the standards you need to follow, such as length, format, due-date, etc.

* If you do a project marked with an asterisk, I'd like to read it. Please send it to me at the address on the back of the title page. Thank you.

Astronomy and Culture
Hollywood and Real Science

Many movies deal with astronomical themes and motifs. Pick one, and tell what aspects were fact, what was plausibly extrapolated from fact, and what was fiction contrary to current understanding. Exploit WWWeb resources. Pick any *Star Wars,* any *Star Trek, Armageddon, Deep Impact, Contact,* or other astronomy-related movies.

Plans for the Next Good
Total Solar Eclipse

Thousands of people will flock to see a good total solar eclipse, "nature's grandest non-violent spectacle." Each country along the path has its own circumstances. Tell about the best places to go, events planned in conjunction with the eclipse, which places are already sold out, etc.

Foreign-Language Sources for
Sky Lore*, Historical Astronomy*,
and Space Programs

Can you read a foreign language well? You can help me by reading sources which are not available in English. Tell about that culture's pre-scientific sky-lore, that culture's historical and current astronomers and observatories, and current astronauts, satellites, and so on. Obtain sources from relatives in the old country, a university library, WWW, or elsewhere.

Astronomical Sites You Travel to

Will you travel during this term? Turn a trip that would otherwise _interrupt_ your course into a vital part _of_ your course. Many places you may visit have astronomically-interesting facilities, such as observatories, planetaria, space bases, and meteorite craters. Research it _long_ before the trip. _Phone ahead_ to arrange a detailed (preferably behind-the-scenes) tour.

Hoagland's Face on Mars*

Compare how scientists analyze imagery of "The Face" to the way pseudoscientists treat it. Include reactions to the _Mars Global Surveyor_ imagery by scientists, by the mainstream media, by pseudoscientists, and by sensationalist media.

Dismantling the Soviet Space Program
for a Russian-reading student

The Soviet Union ran the world's second-biggest space program. It is now crippled, and much is for sale. Get the latest information, including US$ prices for a basic booster, for a tourist ride, cosmonaut advertising, etc.

How Astronomical Institutions Cope in Hard Times

St. Petersburg and Moscow institutions are attacked by Russian thugs who want to sell their land commercially. How badly was Pulkovo Observatory damaged by the fire? How are institutions faring in economically distressed areas?

Astronomical Aspects of Your Major, Job, or Hobby

Many businesses use satellites. Many majors and some hobbies touch on astronomical topics (telescope companies; collecting "Space" stamps ...). These projects benefit greatly from your own expertise, specialty materials, or employer.

Astronomical Evidence in Case Law*

requires affordable help from a lawyer or legal researcher

Drama ... brilliance ... bluster ... disrespect for Science ... Find court cases where astronomers gave expert testimony about astronomical evidence – typically the angle of the Sun, or the brightness of the Moon, but it is especially nice to find other topics, too. Describe the situation, the testimony, how the opposing lawyer treated the astronomer and the expert testimony, and how the astronomer and the expert testimony affected the case's outcome.

Why Kids Learn That The North Star Is The Brightest Star In The Sky, And Straight Overhead, Even Though It Isn't

A huge proportion of schoolchildren coming to American planetaria "know" these things, even though they are not true. How does this happen? Find such children (your local elementary school? Cub Scouts and Brownies? ...) and interview them to find out. Caution: Many children try to tell you what you want to hear, so if you tip them off, you won't get candid responses.

Pseudoastronomy in Tabloid Papers*

Certain supermarket tabloids come up with new wrinkles every few months. For a while horoscopes are hot, then "The Face on Mars", then space-aliens with selective appetites, and so on. What are they up to this month? What's wrong with it scientifically?

Golden Oldies

for voracious readers and history buffs

In parallel with your textbook, read an obsolete or current competing astronomy textbook. Write a "compare and contrast" essay on the 2 books. Keep up with every topic every day since you may be asked in class what the old book said about the current topic. Some old texts are in college libraries, and some are in library storage.

Astronomy and Spaceflight in Popular Culture

How are astronomy and spaceflight handled in: Art, Music, Comic Strips, Advertising, Novels other than Science Fiction, Calendars,

Movies, TV Shows, News Media, Wallpaper, Fabrics, Clothing (T-shirts and ties; anything else?), Tableware, ...

Astronomical Imagery in Magazine Advertisements

Many ads show astronomical imagery. What do they try to convey? Do they commit blunders?

Biographies: astronomers, astronauts or cosmonauts

Space Bases: NASA, IKI, ESA, ISAS or other space agency

Observatories: anywhere, old or new

Planetaria: anywhere, neat shows and exhibits

Scientific Projects
Creating New Astronomical Computer Graphics and Animations

Can you make computer graphics and animations? Here are various novel astronomical demonstrations. Select a project from this wish-list. Research the astronomy. Create the animation or graphics. Well-done versions may be used in future classes.

Telescope focal ratios*: Read my article "Of Pupils and Brightness", *Griffith Observer,* vol. 49, no. 1, January 1985, pp. 14-19; also posted at www.everythingintheuniv.com/telescopes. Develop a standard screen format portraying the cross-section and ray-trace path for types of telescopes of standard 20 cm diameter. For each focal ratio, changing in whole numbers from f/3 to f/20, show what each of the following would look like: achromatic doublet refractor; and reflectors of Newtonian, Cassegrain, Coudé, Ritchey-Chrétien, Gregorian, Dall-Kirkham, Schmidt, Schmidt-Cassegrain, Maksutov, and prime-focus types. Include labeled scale bars in both metric and English units. Include circles showing the size, brightness, and contrast that the human eye would perceive through a 25 mm eyepiece (with 60° field of view) for each of these popular objects: the Moon, Jupiter, the Pleiades cluster (M 45), Orion Nebula (M 42), and the Whirlpool Galaxy (M 51). Users can see that varying the focal ratio dramatically changes image brightness, contrast, and size. A further elaboration would permit eyepiece focal lengths from 4 mm through 50 mm.

Objects Across the Spectrum*: Crossfade imagery of prominent objects in order of wavelength, from every imaging satellite and radio observatory. Each object constitutes a separate project. Certain obvious objects: the Moon, the Sun, Jupiter, the Orion Nebula (M 42), the Crab Nebula (M 1), the center of the Milky Way, the Andromeda Galaxy (M 31), Centaurus A (NGC 5128).

Constellations in 3 Dimensions*: Using new distance data from the *Hipparcos* satellite, find the precise distance to each visible star in a constellation. Animate walk-arounds and walk-throughs which demonstrate their depth. Change brightness with distance. Obvious prominent constellations: Ursa Major, Orion, Coma Berenices, Scorpius, and Cygnus.

Types of Variable Stars*: Develop a standard screen layout featuring a moving graph-line for the star's brightness (and, for pulsating variables, graphs for diameter and surface temperature), labeled with time intervals. Inset a cartoon of the star undergoing its changes. Include the star's name, its type, the speed-up factor (or elapsed time), a thumbnail description of the processes occurring, and the quantity of known members of that type. Where possible, obtain real light-curves from the American Association of Variable Star Observers, 25 Birch Street, Cambridge, MA 02138. Each type is a separate project: R Canum Venaticorum (rapidly rotating close binary stars); Algol (eclipsing binary stars); BY Draconis (rotating spotted stars); T Tauri (unsettled baby stars); Cepheid (regular throbbing red giant stars); RR Lyræ (related to Cepheids, but faster, with overtones); Mira (semi-regular long-period red giant stars); R Coronæ Borealis (dying giant stars coughing out soot); U Geminorum (recurring dwarf novæ). I suggest starting with 10 seconds per cycle but program it so that experience can suggest an optimum speed, and adjusting requires a minor plug-in, not major reprogramming.

The Latest from any:

- Space Shuttle Mission
- Satellite Mission
- Space Probe Mission

Hierarchies

possibly for someone who is very organized

The Universe has a history of getting itself organized. It builds up atoms and molecules, coalesces objects, and distills and stratifies them. This is why the cosmos is not a homogeneous glop. From lecture and textbook, identify all the hierarchies and systems encountered. Draw parallels and overviews of cosmic organization where possible. This is the 'other side' of the Second Law of Thermodynamics (Entropy).

Deep Astrophotography*

This exercise is *only* appropriate for experienced astrophotographers. More things can go wrong in astrophotography than in any other aspect of novice astronomy, so, to avoid frustration, don't try this unless you've already taken astrophotos that win oohs and aahs at astronomy club meetings.

Imaging technology is rapidly improving and cheapening. From current astrophotography guides, determine "recommended" exposures for accessible deep-sky objects. Then take time exposures that are many times longer than the recommended amount – up to the sky-fogging limit. That should grossly overexpose the inner parts, but what will you discover beyond them? This may reveal new structures, some of which have different shapes than the brighter components. Some may show nothing new – itself noteworthy. And all the ones that *do* show something new are worth drawing scholarly attention to. Please send me a copy of any interesting results.

Projects That Address How Science Works

by Brad Schaefer

Astronomy Article Review

Throughout your lives, you will read many articles in popular media that discuss astronomy and related topics. But all research scientists know that articles dealing with their own area of expertise frequently make (sometimes grievous) errors in detail or overview. Also, the new result is often hidden away without enough information to tell how valid the discovery is. And the choice of the specific research item might tell you more about the discoverer's press agent than about new astronomy results.

Select and read some article relating to astronomy in the *popular* media, research this report in the *scientific* press, and report your findings.

The popular media often have articles which discuss astronomy topics. Good sources are the *New York Times* (in their Tuesday Science Times section), or *Time* or *Newsweek* magazines. Ask your instructor about what topics are acceptable; many take any topic covered in the course. If you need help finding articles, or determining how suitable they are, ask your instructor.

Your research should include reading 2 or more articles in the scientific press. This includes *Scientific American, Science News, Sky & Telescope, Physics Today, Mercury,* and *Astronomy.* All these magazines are widely available, well indexed, accurate, and reliable in their Science reporting. Read at least 2 articles discussing the topic of your popular press article. The technical details may be beyond your knowledge, but look for clues to reliability and importance of the new result. You might have to look through several years of magazines to find articles that cover the subject, but the articles themselves are usually fairly short.

Address these questions in your writeup:
(1) What is the article title and citation that you are reviewing? (Please attach a photocopy of the article to your review.)
(2) Give the title and citation of all articles you researched to write your review.
(3) What is the specific new result claimed in the article?
(4) Is this really a new result? That is, was the new result totally unexpected, was it already anticipated, was it a better measure of a known result, or is it just one of a series of similar claims?
(5) What is the dominant source of uncertainty in the reported result? Did the reporter tell the readers about it?
(6) Did the reporter present the basic evidence or logic used to justify the new result?
(7) Did the reporter get the facts and concepts right?
(8) Did the reporter place the result in context? That is, was the new result given its place in the history of the field, is the importance expressed, and were the implications stated?
(9) Is there a controversy associated with the reported result, and if so, why?
(10) Estimate the probability that the new result is right.

Observing Report

For this observing report, carry out an observing program of your own choosing and analyze the data obtained.

The format of your written report must be acceptable to your instructor. Describe your goals, methods, details of circumstances, details of calculations for analysis, a succinct summary of your results, and suggestions on how things could be done better. Attach your raw data in whatever form is appropriate. Be mindful of your accuracy of measurements, estimation of uncertainties, obtaining enough data to solidly achieve the goal, and ingenuity in overcoming problems.

The observing projects all require clear skies on multiple nights, often spread out over a long time. Therefore, procrastination until the last minute will produce a disaster. Start early.

Here are details for 5 sample projects.

PROJECT IDEA #1: ALGOL

GOAL: Construct a graph of changing brightness over time ("light curve") and deduce properties of Algol, a bright eclipsing binary star in Perseus.

OUTLINE: First, learn to locate Algol. Use a finder chart (such as appears near the center of *Sky & Telescope* every month). Then get magnitudes for nearby comparison stars between magnitudes 2.0 and 4.0 – learning the magnitude scale if you haven't already done so. Either find a prepared chart (available in many books), or look up the brightnesses of specific nearby stars in books in the library. Then, carefully compare the apparent brightness of Algol to the various comparison stars to estimate Algol's magnitude. Record both time and magnitude. Eclipses occur at times that can be predicted to a few minutes, as published in the *Observer's Handbook* and the Calendar Notes in *Sky & Telescope* (online at http://www.skypub.com/whatsup/algol.html). During eclipses, estimate brightness every 15 minutes.

MINIMUM: Get a light curve for 10 nights that include 3 times when Algol is in eclipse (with at least one night where you have a complete light curve throughout eclipse). For your analysis, determine the orbital period from your data, and estimate the radius of the faint star.

EXTRAS: Measure the depth of the secondary minimum. Is the orbit eccentric?   Is the eclipse partial or total?   Measure the ratios of brightness of the 2 stars. Observe the eclipsing binary β Lyræ in the same manner.

PROJECT IDEA #2: SUNSPOTS

GOAL: To measure the rotation period of the Sun.

OUTLINE: First, use a **safe** means of looking at the Sun. Never look directly at the Sun without a safe filter (these include welder's glass #14 and aluminized mylar; see B. R. Chou, *Sky & Telescope,* February 1998, page 36). Alternatively, a view through binoculars or a telescope with a safe filter is OK. For this, use only filters sold specifically for Sun viewing that are to be placed over the aperture, and cover any finder scope to prevent inadvertent views.   A third alternative, projection through binoculars onto a sheet of paper, is quick, convenient, and can easily reveal spots. (Use a telescope if that's easy enough.) Never look directly through optics being used for projection. Even with telescopes and with clear skies and with the Sun at zenith and with no filter, it takes longer than 10 seconds for an unwavering stare at the Sun to produce permanent eye damage (Sadun et al. 1984, *Archives of Ophthalmology,* v. 102, p. 1510).

When the Sun is near solar maximum [as in 2001], there is an average of one sunspot group visible to the average human eye without optical aid. Practice to get to the point where groups can be readily seen. If you are not seeing sunspots without optical aid, then to complete this project, you will have to use binoculars [or even opera glasses] in projection to collect enough usable data.

Draw the positions of individual sunspot groups for all clear days for a month. Be careful to mark North uniformly on all drawings. Analyze how long it takes for a group to travel from one edge to the other. That's half a rotation. Note that foreshortening near the edge will slow the apparent motion of a sunspot group.

MINIMUM: Make a drawing every clear day for 3 weeks, and deduce the rotation period of the Sun. To see small sunspots, you might have to use binoculars with the safe projection method.

EXTRAS: Use the drawings to determine the latitude of each group, and also to see if the rotation period changes with latitude. What is the average lifetime of a sunspot group? Where does the Sun's pole point with respect to the ecliptic? Investigate how Galileo used the motion of sunspots to provide his only valid proof of the Copernican System.

PROJECT IDEA #3: δ CEPHEI

GOAL: δ Cephei is the prototype of the Cepheid variable stars. It is easy to see it vary in brightness. Construct a graph of changing brightness over time ("light curve") and deduce the distance to the star.

OUTLINE: First, learn to locate δ Cephei. Use a finder chart (such as appears near the center of *Sky & Telescope* every month). Then get magnitudes for nearby comparison stars between magnitudes 3.5 and 5.0 – learning the magnitude scale if you haven't already done so. For this, you'll have to either find a prepared chart (available in many books) or look up the brightnesses of specific nearby stars in books in the library. Fortunately, 2 very nearby stars (ζ and ε Cephei) are adequate comparisons. Then, carefully compare the apparent brightness of δ Cephei to the comparison stars to estimate its magnitude. Record both time and magnitude. A good light curve can be gotten from hourly estimates made on 10 nights. The star at its faintest is difficult to see from light-polluted cities, so a dark site or binoculars would help a lot.

MINIMUM: Get a light curve with 25 estimates spread out over many nights. Then use your data to find the period of variation and the average magnitude of δ Cephei. Then, use this information (along with a Cepheid period-luminosity relation) to determine the distance to δ Cephei.

EXTRAS: What is the distance to δ Cephei as based on your measures? Is this too far for a parallax to δ Cephei to be measurable? Observe η Aquilæ in the same manner.

PROJECT IDEA #4: LIMITING MAGNITUDE

GOAL: For a group of at least a dozen friends, measure how faint a star they can see, and at what magnitude they switch from direct to averted vision.

OUTLINE: First, copy appropriate star charts for the evening sky. A source would be *Sky & Telescope*. Then gather a large group of friends into a dark site somewhere near campus on a clear moonless night. The people must get totally dark adapted for at least 15 minutes. Each person should use a dim, reddened flashlight and the star chart to locate key stars overhead. [Go out several times in advance to learn the constellations really well, otherwise your friends will get annoyed and frustrated. Also, bribes of hot pizza, or such, should help.] Then, specify a large section of the sky, in which they should mark down every star which they see, and whether they can see it by staring directly at it or must look away with "averted vision". Each person should record a dozen stars that they can barely see, and mark a dozen stars the chart shows that are invisible. This project must be done with all observers under one uniform sky condition. Back indoors, tabulate the magnitudes of all stars that each observer saw (with direct vision and with averted vision) and didn't see. From this list, deduce the limiting magnitude for each observer, and the magnitude where direct and averted vision switch over. Take the limiting magnitude to be that magnitude for which the observer has a 50% probability of spotting a star.

MINIMUM: At least a dozen observers where each has recorded the identities of at least 4 stars within one magnitude of their limits. The analysis consists of deriving the limiting magnitude and magnitude where direct and averted vision switch over, for each person. What is the relation between these 2 numbers? How much scatter is there from person to person?

EXTRAS: Brad Schaefer is highly interested in these data, using them for his work on the brightnesses of historical supernovæ and the Hubble Constant. 2 articles you might want to check out are Schaefer, 1996, *Astrophysical Journal,* v 459, pp. 438-454, and Schaefer, 1996, *Astronomical Journal,* v 111, pp. 1668-1674. What implications do your data have for the size of our Universe?

PROJECT IDEA #5: METEORS (optical and radio)

GOAL: To measure the variation in the rate of meteors with time of day, and deduce the average orbital velocity of the meteoroids. Alternatively, meteor showers can be tracked, and the width of the debris swarm measured.

OUTLINE: Meteors are easy to see. Just lie on your back under a clear dark sky and wait for the flash. To see substantial numbers, you must get to dark skies. Alternatively, you can tune your FM radio to a station 700-2100 km distant and listen for short moments of signal caused by the radio waves bouncing off ionized meteor trails. With this method, you can monitor meteor rates even in the daytime. For full details, see *Sky & Telescope* for December 1997, page 108. Meteor rates vary with time of day, and day of year. Rates are generally highest during the morning and lowest in the evening, for the same reason that rain hits the forward windshield of a moving car while the back window gets little rain. Rates are also high during particular times of the year when meteor showers pelt the Earth with debris left behind by disintegrating comets. The rate variations can be measured by simply counting the number of observed meteors per hour. Items to measure are either the ratio of the morning-to-evening rate or the duration of the meteor shower.

The analysis requires a knowledge of the Earth's orbital velocity, which can be gotten by realizing that the Earth travels nearly in a circle of radius 1 AU in a time of 1 year. Assuming that meteors come from all directions, estimate the volume swept up by the leading edge of the Earth versus the trailing edge as a function of the average meteor velocity. A comparison with the observed ratio will reveal the typical orbital velocity of meteors. For meteor showers, their duration can be translated into a width of the debris stream with the velocity of the Earth.

MINIMUM: For the daily variations, record at least 4 hours at each of 3 times of day, and deduce the average meteoroid orbital velocity. For the meteor shower, record at least 12 hours of observations spread out over all available nights of the shower, and plot the rates.

EXTRAS: For visual observations, estimate the typical angular velocity of a meteor (in degrees per second) and deduce its velocity for an assumed altitude of 100 km. For visual observations, make drawings of the meteor paths on star charts and determine the radiant. For radio observations, compare the rate for stations at different azimuths and explain the difference.

Other Ideas

You can also think up your own projects, perhaps based on your own personal interests or some old burning question. Here are sketches of several more ideas:

(1) Collect a long series of sunset timings where the Sun sets/rises into a water horizon. Look for Green Flashes. Explain in detail how the Sun's declination affects this time, and compare with observations. Look for variations around the predicted times due to thermal inversion layers. This study has applications to archæoastronomy, Islamic prayer practices, and accident law suits [see *Sky & Telescope* 1989, vol. 77, pp. 311-313, and *Science News* 1990, vol. 138, pp. 236-237.]

(2)   With help from observations made by a distant friend, use Eratosthenes' method to measure the size of the Earth; also measure the latitude of this college campus; and measure the longitude here with respect to some distant city.

(3)   Archæoastronomy pursues the paradigm that ancient sites incorporated orientations towards significant astronomical directions along the horizon. So for example, the main avenue of Stonehenge shows midsummer sunrise over the Heel Stone. A perpetual problem is that such alignments might occur by chance. As a control study, discover several significant 'alignments' on campus, where something points accurately towards the solstice sunrises or sunsets.

Index

QUICK ORDER FORM

eMail to **wonttell@rcn.com**
Postal: **Everything in the Universe**
413 Poinsettia Avenue
San Mateo, CA 94403

Please send these:

Quantity	Item	Each	Total

Books

| | *What Your Astronomy Textbook Won't Tell You* | $24.95 | $____ |
| | *Starhopper's Guide to Messier Objects* (23 pp.) | $8.95 | $____ |

Giant 38 x 26-inch AstroMural Posters

	Andromeda Galaxy, M 31(see pp.148, 150, 152)	$9.95	$____
	Horsehead Nebula (see pp. 139-140)	$9.95	$____
	Orion Nebula, M 42 (see p. 138)	$9.95	$____
	Solar Prominence (see p. 106)	$9.95	$____
	Total Solar Eclipse Corona (see p. 63)	$9.95	$____
	Venus Hadley Vortex (see p. 78)	$9.95	$____
	Whirlpool Galaxy, M 51 (see pp. 120, 150)	$9.95	$____

40% discount for 10 or more of each book, or any 10 posters.
Please add <u>sales tax</u> if shipped to a California address.
<u>Shipping in US:</u> This book: $4 for first book + $2 for each additional.
　　　　Starhopper: $2 for first book + $1 for each additional.
　　　　Posters: $4 for first poster + $1 for each additional.

Please make <u>checks</u> payable to **Everything in the Universe.**
<u>Credit cards:</u> Visa or Master Card.
Card number: _____

Expires: _____ Signature: _____

Send to: Name: _____

Address: _____

City, State, ZIP: _____

Telephone: _____ eMail: _____